INSIDE SCIENCE

INSIDE SCIENCE

Stories from the Field in Human and Animal Science

ROBERT E. KOHLER

The University of Chicago Press
Chicago and London

The University of Chicago Press, Chicago 60637
The University of Chicago Press, Ltd., London

For more information, contact the University of Chicago Press, 1427 E. 60th St., Chicago, IL 60637.
Published 2019
Printed in the United States of America

28 27 26 25 24 23 22 21 20 19   1 2 3 4 5

ISBN-13: 978-0-226-61798-5 (cloth)
ISBN-13: 978-0-226-61803-6 (e-book)
DOI: https://doi.org/10.7208/chicago/9780226618036.001.0001

Library of Congress Cataloging-in-Publication Data

Names: Kohler, Robert E., author.
Title: Inside science: stories from the field in human and animal science / Robert E. Kohler.
Description: Chicago: The University of Chicago Press, 2019. | Includes bibliographical references and index.
Identifiers: LCCN 2018036562 | ISBN 9780226617985 (cloth: alk. paper) | ISBN 9780226618036 (e-book)
Subjects: LCSH: Life sciences—Fieldwork—Case studies. | Life sciences—Fieldwork—United States. | Nature observation—Case studies. | Participant observation—Case studies. | Nature observation. | Life scientists.
Classification: LCC QH318.5 .K64 2019 | DDC 570—dc23
LC record available at https://lccn.loc.gov/2018036562

♾ This paper meets the requirements of ANSI/NISO Z39.48-1992 (Permanence of Paper).

For my readers

CONTENTS

CHAPTER 1

SITUATING

In nature and in human societies, the course of every action and event depends on the context in which it happens: that is the premise of this book. And its thesis is that scientific investigations of these activities are likewise best carried out in context. How could anyone think otherwise, one might ask, when the force of situations is a fact of daily experience? Yet in the culture of modern science—at least at its high end—true knowledge is taken to be what is true everywhere, in any context, or in no context. So the scientific procedures considered exemplary—experiment, hypothetico-deduction, randomized trials, statistical proofs of two-variable cause and effect—are those that eliminate or define away contexts as extraneous complications. A sharp social and epistemic boundary is thus drawn between the methods of science and the customary ways of everyday life. *Desituating* has become a signature characteristic of modern science, while science that does not control for context but accepts it is regarded as less than best practice.[1]

These are, of course, ideals. In reality, much science does not evade or suppress situation in a quest for laws that transcend all particulars of place and time. Situating practices are common in the human and social sciences; and in field and environmental sciences, investigations in situ are no less valued than those performed ex situ, in labs and offices. The one creates conceptual space and investigative opportunities for the other, and mixed modes are not uncommon. The more one looks for science in situ, the more one finds. Situating, arguably, is no less characteristic of modern science than its opposite: just less visible and less valued. It is seen as the preliminary groundwork for real science, assembling useful facts but yielding no proofs: that is the modern default view. Part of my purpose in writing this book is to make science in situ a bit more visible, and to expose the disconnect between the powerful but restrictive ideals of desituating science and the reality of creative achievement in sciences carried out in place. I write from the context side, and for it:

as a humanist historian whose business it is to recover the forgotten or misremembered contexts of historical actions and events, I have a brief to make for context.

My particular subject here is an intensive kind of situated science that I call *resident* science. The paradigmatic and most familiar resident practice is the "participant observation" of social anthropologists, who reside in a community for months or years, observing its customs and culture and taking part in a limited way in its activities. Participation is not a necessary element of resident science, however. In studies of animal communities, for example, observers cannot, or should not, take part. A virtue of resident science as a historical category is that it treats a range of human and natural sciences together that are customarily treated separately. And though it is a small subset of the more capacious and diverse category of situated science, resident science epitomizes the defining character of that larger group.[2] I chose it for my subject because it is a well-defined and bounded topic and a manageable one for a shortish book on a very large subject.

What, then, are the defining features of resident science? One is its reliance on intensive personal observing in situ, by observers who in effect operate *inside* their objects of study. Resident science is strongly observational, and often open-ended and exploratory. It seeks generalities in patterns of observed particulars more than in deductions from abstract "laws" and theories. Resident observers treat the contexts and situations in which they and their subjects act, not as stage settings for actions, but as essential elements of phenomena. Resident science is *co*resident. It is situat*ed*, in that observers are themselves present in the situations and actions they observe. And it is situat*ing*, in that subjects are observed in the natural or social contexts in which they normally act. This double situatedness of cohabiting observers and observed is the essence of resident science as I understand it.

My current venture into resident science builds on the extensive inquiries into place and practice in science that have in the last thirty years become a defining issue in science studies. The principle themes of this body of scholarship are now widely familiar: that places are not neutral stages for scientific activities but directly affect how they are carried out and received; that all scientific knowledge is initially the product of some particular material and social locale; that some places—"truth spots"—can give credibility to knowledge claims; and that in becoming what everyone knows—the universal view from nowhere—knowledge must travel and transcend locality of place.[3] These issues of place are important and fruitful. They are not my subject here, however. I am concerned solely with the relations of scientists and their subjects in their common contexts of coresidence: how these working relations are created and how they operate (or don't). Whereas concepts of place have typically emphasized geographical or spatiocultural features, residence is less a spatial than a relational and ecological instrument of thought. So I have kept the more familiar physical issues of place in the background and focused on the activities and experience of investigation—typically of human or animal communities—from the inside, and how coresidence leads not to the erasure of contexts but to their persistence in the facts and concepts of the science produced. Resident science stays in context and keeps context in. Its best stories are inside stories.

SITUATING SCIENCE

To see how situating is achieved, it is useful to look first at its more familiar opposite. *De*situating has been accomplished by various means. Most simply (and superficially) it is achieved *rhetorically*, by stripping written accounts of situating particulars of who and where, in order to create the impression of timeless universality—of Nature speaking, not I or we. This is the process famously described by Bruno Latour and Steven Woolgar in their landmark book on making facts in laboratory science.[4] More substantively, desituating is achieved *conceptually*, by embedding things or phenomena in an abstract framework of theory or modeling. Subjects may also be *materially* desituated, by physically relocating them to sequestered environments designed for the sole purpose of scientific analysis, in which disturbing particularities of context can be controlled or (more or less) eliminated. Laboratories, museums, gardens, aquaria, clinics, field stations, workshops, and offices are familiar types. This *actual* desituating is the favored method of modern science.

The iconic images of desituated nature are familiar: rats in mazes, small animals in cages, rows of maize in weeded and watered plots, fruit flies in jars, pure-strain bacteria in petri dishes; monkeys in vivaria or backyard caged colonies.[5] In the human sciences there are native informants answering questionnaires on an ethnographer's veranda, kids in classrooms watched through one-way glass, or college undergraduates in psychology experiments—perhaps the closest thing there is to a "standard" human organism. And there are the objects extracted from their natural or social contexts: animal skins in cabinet drawers, dried plants pressed in folios, archaeological and ethnographic artifacts in cabinets, and excised tissues and organs in vitro that stand in for intact organisms in vivo. And let us not forget the models of population cycles that may or may not match the real things, or the reassuringly smooth curves of neoclassical economics—stand-ins for economies that in reality are given to "irrational" (yet perfectly natural) swerves, swoops, and dives.

The practical and epistemic advantages of desituating are clear enough. Things and phenomena are more precisely categorized and measured without the complications of real-world contexts, and more securely treated statistically and experimentally. Hypotheses of cause and effect are more easily generated and tested when single variables are isolated and analyzed without the uncertainties of particular situations. Observations and experiments that are reliably reproducible are more easily credited as true than are those that vary with circumstance. Although the noise of context is inherently part of the scientific signal, noisy signals are harder to sell as true knowledge. So science that is presented as clear signal without the context enjoys an epistemic free ride, while situated science must be continually justified and defended.

The social advantages of desituating are no less vital, given a society where professional reputation and advancement in a career depend on regular publications produced by communally sanctioned procedures. Desituating practices facilitate the steady production of results, enabling practitioners to publish and not perish; and because they stand up to critique and peer review, they are more effective in making and defending reputations.[6] Situating practices are in contrast less reliably productive, and because they are carried out in person and in particular contexts, they will likely be harder to represent as objective. Results that are not presentable as definitive answers to specific hypotheses will be harder to get funded and published. It is little wonder, then, that desituating sciences are high on the totem pole of scientific prestige, while those that embrace situation are lower down.

All this is not to say that situating science has been abandoned in modernity: it obviously has not. Natural history has lost standing in the age of experiment, but it may well be pursued by more people and more productively now than ever before. In ecology and environmental sciences—the modern descendants of natural history—in situ observing is common practice. In some sciences it defines the mainstream, as in social anthropology and wildlife ecology; in others it animates significant subfields, as in animal behavior and small-group sociology and human relations. In many sciences recurring efforts (not always sustained) are made to resituate abstracted objects of study, as in evolutionary biology (see below) or social psychology, and even a science as devotedly abstract as economics.

There are good reasons why, despite their practical and epistemic disabilities, situating practices have retained a significant place in modern science. They restore contextual meanings that are too easily sacrificed to single-minded pursuit of analytic rigor. They keep scientists mindful of the fact that contingency and uncertainty are essential aspects of phenomena, and that analytic categories and cause-and-effect connections may not be so securely real as they seem to be when identified in unnatural isolation.[7] Open-ended exploration revitalizes domains of theory that with time and use have become merely self-confirming, and enlarges conceptions of "data" that have become narrowed to just what can be used in rigorous "proofs" of cause and effect. Intensive study of particular cases in context challenges a methodology of deductive hypothesis testing that becomes unhealthily hegemonic, as the political scientist John Gerring has observed.[8] Situating is a useful reminder that "just because an idea is true doesn't mean it can be proved. And just because an idea can be proved doesn't mean it's true."[9] Observing phenomena in context is, in sum, not an imperfect substitute for proper science: in many instances it *is* proper science. Abstraction and universalism are not in themselves the problem: only the assumption that they are the best method for every subject and situation. There seems to be a growing recognition in science and in history of science that universal "laws" may not be the best way, and are certainly not the only way, to deal with complex phenomena.[10] One size does not fit all.

Those who have explored the social and epistemic advantages of decontextualizing have generally not symmetrically inquired what is lost when phenomena are stripped of context. In his pathbreaking essay on Louis Pasteur's anthrax field trials, for example, Bruno Latour demonstrates how Pasteur captured anthrax for laboratory medicine by relocating the phenomenon from a pasture, where infection occurred contingently and mysteriously (it seemed), and farmers and veterinaries were the authorities; to a laboratory from which contingency was banished, and where infection was made reliably repeatable and explicable as the effect of a certain species of microbe. That indisputable cause and effect gave bacteriologists effective ownership of the anthrax disease: the one road to truth passed through their labs.[11] It remained nonetheless true, however, that in the real world of herds and meadows, infection depends crucially on the particulars of environment and circumstance. And these cannot be properly studied by Pastorians in labs but only in pastures by those who know them and know how nature and human husbandry work. The question here is, How are objects of science kept in their pastures, so to speak, or returned to them from Paris labs for scientific study in situ?

There are degrees and varieties of situating practices. Rhetorical situating—I was there, and here's what I saw and did, believe me—was once the usual form of natural histories and is still occasionally used to create a sense of virtual witnessing.[12] Situating is also accomplished by expanding the conceptual frames in which phenomena are set: genes reset in the whole machinery of gene expression, cells in nets of cell communication, microbial species in microbiomes of vast diversity, plant and animal species (including the human species) in whole ecosystems. A striking recent case is the resituating of the mathematical genetics of altruism and eusociality (think ants and termites, or us) into the larger frame of ecology and evolution by Edward O. Wilson and others. It is not just any species that develops eusocial traits, it turns out, but those few that have chanced to evolve ecologically to a point where eusocial tendencies can be amplified by natural selection. "[T]he origin of altruism cannot be deduced by aprioristic reasoning based on general models," Wilson writes. "It can, however, be revealed by reconstructing actual histories with empirical data."[13]

Abstract models may similarly be given richer conceptual surrounds. Theoretical economics affords a good example in the logical game of prisoner's dilemma, in which two idealized "rational actors" kept incommunicado weigh the benefits and costs of cooperation or betrayal. This universalized model, admitted by all to be totally unrealistic, is made (modestly) more true to life by adding little stories with more actors and interactions, and particulars of social context. A reductive logic of decision making is thus resituated in imagined but plausible scenes of choice.[14] In principle a story could be invented for any actual situation, though that would in the end transform a *reductio ad absurdum* into an *elaboratio ad absurdum*. One is reminded of Jorges Luis Borges's fable of an ancient college of cartographers, who created a map of their empire as big as the empire itself—a perfect but perfectly useless object that was finally abandoned, along with the college and all cartographers.[15]

More concretely, the situations of experimental subjects can be made materially more like those of nature or society. *Mimetic* situating, one might call such practices. Minimally, "amenities" (exercise wheels, natural nesting materials) may be added to animal cages in the hope of making captives' behavior more like what it would be in free life. Vivaria and aquaria are made more naturalistic by adding species, or environments that afford cover from predators. Entire laboratories may be located in places whose climates and environments cannot be simulated: alpine labs for thin-air physiology, marine stations for littoral and intertidal environments, arctic stations for cold-climate science, or gardens in various climate zones for study of plant variation. Or actual situations in nature may be discovered or arranged that do in context what experiments do in labs—*nature's* experiments and what I've called "practices of place."[16]

Alternatively, objects extracted from their natural environments may be delivered to museums or herbaria together with context in the form of field notes detailing when and where objects were collected and from what natural or social situations: little stories of context that are not imagined, now, but real. Specimens may be accompanied by bits of material context for dioramas that mimic collecting sites. Museum curators and scientists resituated themselves when, rather than delegating fieldwork to amateurs and commercial collectors, they went afield to do their own gathering and observing, typically in seasonal expeditions.[17] Epidemiologists, after the first flush of petri-dish bacteriology, likewise went afield to observe how infections worked in actual situations, thereby making microbiology a kind of human ecology.[18] Recurring or *commuting* situating, we might call these expeditionary practices.

But the most thoroughly situating practice is residing for extended periods among subjects of study. *Resident* observing is not one practice but a set of practices that vary with particular subjects and observers' aims and situations. Resident observers may become limited participants in their subjects' daily activities in order to experience for themselves ways of life not their own—the standard practice of anthropologists. Or they may take care not to involve themselves with subjects that do not tolerate participation, as ecologists and ethologists do with free-living animals. Residence is thus not mere physical propinquity but an active relationship between observer and observed in contexts of coresidence, and the relations of cohabitating are my chief concern here. That is why I generally prefer the term *situating* to *situated*. The past tense "situated" invites us to regard situation as a set stage for actions. The present participle "situating" points to the actions themselves and highlights the fact that residing is a creative practice, transforming situations of life into those of science, and vice versa. In resident sciences the practices and experiences of situating are most fully and openly on display.

I would not want to leave readers thinking that desituating and situating science are dichotomous and incommensurable. I have dwelt on their differences to highlight the less familiar qualities of situating science. In fact, the two are often compatible and complementary. Thomas Gieryn, for example, has shown how the sociologists of the Chicago school drew on the epistemic advantages of both lab and field to justify their use of city neighborhoods as research sites, shuttling with little sense of incongruity between the situating values of *found*, *here*, and *immersed* and the desituating values of *made*, *anywhere*, and *detached*. Chicago was for them simultaneously a laboratory of social science and a field of human ecology.[19]

There is still, to be sure, the ingrained modern bias favoring practices that desituate and universalize over those that situate and deal in particulars. Theoretical deduction and experiment need no justification; observing and describing often do. That has been a plain fact of life and a subject of rueful complaint for many who have pursued science in situ, and it would be unhistorical to deny it. Yet that epistemic disadvantage has not prevented scientists from drawing on situating and observational practices if these do the job. To once more quote John Gerring, "[M]ethods, strictly defined, tend to lose their shape as one looks closer at their innards. . . . There are few 'pure' methods. And this is probably a good thing. Chastity is not necessarily an attribute to be cherished in research design."[20] Impurity of practice is arguably one of the chief strengths and virtues of situating sciences. These constitute border zones in which varied scientific practices, together with those of everyday occupations, can be promiscuously and productively mixed.[21]

I examine in this book six cases of resident science. The ethnographic practice of participant observation is the first, exemplified by Bronislaw Malinowski's sojourn in the Trobriand Islands of Melanesia (1916–19). My second and third cases are two resident investigations of modern societies by sociologists: Nels Anderson's, of itinerant hobos in Chicago's so-called Hobohemia (1921–23); and William F. Whyte's, of Italian-American street-corner cliques of young males in Boston's North End (1937–40). Primatology is my fourth case, represented by Jane Goodall's long-term study of chimp behavior in the Gombe Stream Chimpanzee Reserve of Tanzania (1960s and beyond). Finally, there are a pair of studies in wildlife ecology: Herbert Stoddard's pioneering life history of bobwhite quail in the longleaf pineland of upland Georgia (1924–29), and the study of quail ecology and predation by Paul Errington and Aldo Leopold in south-central Wisconsin (1932–40s).

In selecting these cases, I was guided by both programmatic and personal concerns. I came to the subject of resident science somewhat serendipitously via a study of wildlife ecology, which I had thought was typical but realized was not. If I had come to the subject more deliberately, I asked myself, with what science would I have begun? Anthropology and Malinowski was the obvious answer; and that led easily to the sibling science of sociology and the Chicago school. To bridge the species and historiographical divide between human and animal science, the choice of primatology and Jane Goodall was an easy one. And for a concluding case, there was the one that had serendipitously opened the larger question of resident science, now substantially reconceptualized.[22] I also had practical and personal reasons for choosing these particular cases. Each is richly documented in scientific publications, biographies, and autobiographies: a necessity, as I had determined not to use archival sources (the subjects are too many and too diverse to dig that deeply into each one). I also wanted individuals who would be engaging personal and intellectual company for the years it would take to learn about subjects of which I then had but a passing knowledge. I wanted good stories—inside stories—to tell of my actors' lives and their experiences with resident observing, which would be enjoyable both for me and for readers who (like me) are curious about all kinds of science but not aspiring specialists in any one.

It is obvious that my cases of resident science do not constitute a natural group, in the sense of having common origins or interconnections, or of marking a historical trend. They do constitute—by design—a historiographical series from a fully representative case of resident practice to one that might arguably be a limiting case. Having no hypothesis to test, I do not ask the same questions of each subject, but rather follow lines of inquiry that seem most interesting and apt. My approach, like those of my subjects, is analogical and exploratory. These are the qualities of case study generally.[23] As the sociologist Howard Becker has forcefully argued, the best use of case study in the social sciences (and, I would add, in history) is not to produce universal "laws" or generalizations but to discover features of social phenomena that analytic methods of sampling and hypothesis testing do not. Becker reasons analogically: finding that some variable plays a significant role in one case, we should reflexively ask if it might also be significant in others, where there was no obvious reason to look for it. By this freestyle cross-examining of cases, vital issues of social life may be brought to light: "What reasoning by analogy from known cases gets you isn't guaranteed knowledge about an entire population of similar cases but something more valuable and less ephemeral: a collection of connected researchable questions about a family of related phenomena, ideas that can orient your next research from the first moments of aroused interest through the actual research."[24] Becker regards case study not as an empirical preliminary to formal analysis of cause and effect (the common view), but as an exploration that enriches familiar categories and discovers new ones. It is a method not of abstraction and universals but of patterns discovered in empirical particulars.

CLUES AND CONJECTURES. The historian Carlo Ginzburg sets forth a similar vision of scientific and historical method in his brilliantly idiosyncratic 1980 essay on "the method of clues."[25] Ginzburg means by that phrase the practice of inferring from seemingly trivial observations the unseen realities of past or hidden actions. Sherlock Holmes is the iconic practitioner of the method of clues—the things that should be there but are not (barking dogs in the night), the unnoticed leavings that betray even the most careful criminals to the detective who sees everything and has a deep knowledge of the world of material stuff and human behaviors. It is a situated practice of observing inside a natural or social world. It's how we navigate our own ways in the world: hence, no doubt, the undying appeal of detective stories.

Ginzburg's exemplars of the method of clues are strikingly diverse. Art historians and connoisseurs assign authorship not by the overall look of a work of art (too easily faked) but by common small details like fingers or ears, which, because they are unthinkingly produced, are fingerprints of identity. Freudian psychoanalysts "divine secret and concealed things from despised or unnoticed features, from the rubbish-heap, as it were, of our observation." Medical diagnosticians observe the unobservable—diseases—from signs that to the inexperienced eye, ear, nose, or hand are meaningless. The origins of the method of clues, Ginzburg notes, lie deep in human history—in practices of divination and astrology, which read the unobservable minds of gods from arcane details of entrails, thrown sticks, or planetary alignments. And deeper still, "one glimpses something as old as the human race: the hunter crouched in the mud, examining a quarry's tracks." From the slightest physical signs—paw prints or trails, snapped twigs, faint smells, scat, bits of fur or feather—hunters read the presence of unseen prey or predators and the causes of unseen acts.[26]

In modern science we see the method of clues in Georges Cuvier's famous trick of conjuring up extinct animals from the morphology of a single bone; and in the method by which Charles Darwin, with no hope of finding smoking-gun evidence of speciation by natural selection, argued the case from innumerable small details of natural history. The "method of Zadig," Thomas Henry Huxley called it, referring to Voltaire's version of the medieval Persian story of three men who correctly described a lost camel from visible traces, though they had never seen the animal itself. The method of clues is also the mainstay of historians, who must reconstruct past actions and events from fragmentary empirical records, as well as of historical sciences like archaeology, geology, and paleontology, and of the human and social sciences, which seek the origins of customs and institutions in myths and memories.[27] To Ginzburg's list I would add ecology and systematics, which likewise look to small details to distinguish between different species and kinds of ecosystems; ethology and wildlife science, which read meanings from small clues of context and behavior; and, perhaps, resident science generally.

Ginzburg's method of empirical clues is also a method of interpretive *conjecture*. In its modern usage, "conjecture" has connotations of careless or ill-founded reasoning, like "surmise" or even "guess." The implied contrast is with the abstract reasoning of mathematics, idealized models, and experiment—what Ginzburg calls (with no disrespect to the historical man) the "method of Galileo." That was not the original meaning of the word, however. In its premodern usage, the implied contrast of "conjecture" was not with a superior human way of knowing but with the perfect and unattainable knowledge of divinity—the word derives from the common practice in divination of "throwing together." Conjecture was the best that mortal man could ever do. Ginzburg in effect revives that older meaning as a sound and serviceable way of understanding a world in which certainty is seldom if ever to be had by any means. It was only the "powerful and terrible tool of abstraction," he writes, that in modernity made conjecture seem an imperfect second best.[28] Conjecture, as John Gerring writes, "is quickly dismissed as a matter of guesswork, inspiration, or luck—a leap of faith, in any case, and hence a poor subject for methodological reflection."[29] Ginzburg mounts a vigorous defense of conjecture in science. The essential qualities of actions and phenomena lie in their particularity and the contingencies of context, so that applying the abstracting "method of Galileo" removes from them precisely what might make their causes and courses understandable. If clues and conjecture lack the perfect but artificial rigor of theoretical deduction, they have an "elastic rigor"—a usefully paradoxical phrase—that is better suited to understanding the actual mixed-up world, because it is a rigor of direct experience of particular phenomena in context.[30]

A SCIENCE OF PARTICULARS. Resident science is science of particulars, though such a thing is commonly thought to be a logical impossibility. Knowledge of particulars is a practical necessity in worldly activities, but the honorific "science" is reserved for universals. Knowledge of particulars is "know-how."[31] This epistemic bias has deep roots. As the philosophers Samuel Gorovitz and Alasdair MacIntyre have argued, it was a principle of Aristotelian philosophy that was quietly retained in the transition from medieval natural philosophy to modern science. And it persists to this day in the familiar dichotomies of "pure" and "applied" science, experiment and description, and science and common sense; as well as in a status hierarchy of sciences that elevates what is abstract and universal and demotes what is situated and particular, and "blinds us to the existence of particulars as proper objects of knowledge."[32] These habits of thought persist despite the obvious fact that the world consists of nothing else but complex particulars, which have histories and contexts too complex and contingent to be understood in terms of universal laws. Gorovitz and MacIntyre take hurricanes and salt marshes as paradigmatic subjects of a science of particulars; but just about anything that has life and agency or that depends on the contingencies of context would also serve—that is, just about everything that is and happens in the world.

Gorovitz and MacIntyre sketch out some of the basic principles of a science of particulars. One is that all things and phenomena are proper subjects of science: messy cases cannot be ruled out as not-science just to save the appearance of universality. A second is that scientific "laws" need not be rigidly law-like, but may accommodate exceptions and counterexamples. Vaccination, for example, does reliably confer immunity, notwithstanding occasional idiopathic failures. A third principle is that history is always a necessary consideration. Omitting it in order to streamline cause-and-effect proofs is unscientific. Likewise with natural or biosocial contexts: these are not noise but signal, and omitting them achieves a rigor that may be illusory. A related point is that a science of particulars is a science of cases—as in common law or clinical medicine. Knowledge of all aspects of a single case may be more revealing than knowledge of a few selected variables in many cases. Gorovitz and MacIntyre suggest further that in a science of particulars, deep experience of actual objects or actions will often be more fruitful than virtuoso skill in formal procedures and methodologies: "What is important to the theoretical or experimental scientist is experience in research, not experience of the distinctive features of the particular crystals or molecules or other entities which provide a subject-matter for the research. . . . [W]hat is important to the meteorologist, navigator, or veterinary surgeon is an understanding of particular, individual hurricanes, cloud formations, or cows, and thus what is distinctive about them as particulars is what is crucially important."[33] It follows that purity of formal procedure should count for less in judging the credibility of scientific products than producers' knowledge of the particulars of their subjects.

Gorovitz and MacIntyre's manifesto for a science of particulars comes in a weak and a strong version. The weak version is a call for methodological pluralism and tolerance: "The dominant interpretation of natural science," they write, "must be revised so as to allow a place for our knowledge of particulars alongside our knowledge of generalizations." Live and let live, both are good in their proper places. No arguing with that. The universalizing methods of Galileo are a problem only when they become hegemonic—the one best way. The strong version of the case for a science of particulars is more contentious and interesting, because it turns the conventional modern order of standing of the sciences radically upside down:

> [I]n this stronger version the thesis would insist that nature consists of nothing but more or less complex particulars, that theoretical physics is the most abstract kind of knowledge, and that it therefore always has to be based on our knowledge of particulars gained by means of sciences of the concrete. The most fundamental sciences on this view would be the disciplines concerned with our practical transactions with particulars: medicine, veterinary medicine, engineering, military and political sciences, and so on.[34]

The most abstract sciences are, in other words, not foundational to "sciences of the concrete" but parasitic on them. And I would include in

Gorovitz and MacIntyre's category of "concrete" science, along with applied and practical disciplines, much of human and social science, the historical sciences, ecology and environmental sciences, and, potentially, any science that is pursued intensively from the inside.

THEMES

Of the many intriguing features of resident science, a few seem to me of particular interest. One theme that has grown on me in the drafting and redrafting of this book is how scientists' life experiences materially shape not just their choice of subjects and modes of practice but also those practices themselves, as everyday activities and occupations become the means to scientific ends. The two essential practices of resident observing—residing and observing—are amplified forms of common activities, and my second and third themes are how they also operate as scientific practices. Finally, there is the theme of stories and storytelling. A striking feature of my cases of resident science is how practitioners use stories—of their subjects' activities and their own—not just to represent the results of their work, but also as tools of factual inquiry and reasoning. Like living, residing, and observing, storytelling is another everyday activity that becomes a practice of science.

LIFE IS SCIENCE. That the common knowledge of everyday life and the expert knowledge of science are incommensurable domains has been a cardinal principle of modern science, and the logical foundation of its claim to epistemic authority. The casual activities of everyday life and the purposeful and disciplined methods of science are in this view essential in their proper spheres, yet are different modes of experience and meaning, and not to be confused or conflated. The institutions of modern science embody and sustain this view of separate worlds. Scientists almost universally enter into their vocations through the liminal institutions of formal education, where novices leave the imperfect world of "common sense" behind—"the unreflective opinions of ordinary man" is one, tendentious, dictionary definition of that term—and are initiated into the profession's exacting procedures and methodological dos and don'ts. The business of life is one thing; the calling of science, quite another.

The sociologist and philosopher Alfred Schütz has developed this idea of separate worlds into a metaphysical system in which intrusions of everyday activities into science are not just disapproved, but impossible. In everyday life what individuals think and do is determined by personal concerns—pasts and futures, plans and projects, social interactions of friends and family, occupation, and community—their "Here" in Schütz's terminology. When individuals enter into the social world of science, however, what they take to be important and meaningful is determined solely by the stock of accepted knowledge of their science and its communal practices. In that structure of "relevances," or "Here," any everyday thing is irrelevant and out of place—in effect, invisible to science. That being so, scientists have no way of injecting elements of everyday life into the science—except temporarily, to establish working relations with subjects as a preliminary to the science. "The scientific problem, once established, determines alone the structure of relevances," Schütz writes.[35]

Though theory declares it impossible, in fact everyday occupations and activities have often been turned to scientific ends. Such cases are especially to be found in the situated or resident sciences, where everyday material and social realities are powerfully and inescapably present in the contexts or "Here" of scientific work. The illusion of separate worlds that is created in labs by material and social restrictions is harder to sustain in worldly situations. In situations of coresidence, scientific observers will be less likely to feel constrained by procedural niceties, and will be more open to making use of everyday pursuits and occupations if these seem effective means to scientific ends. Coresidence in effect reunites what philosophy and professional institutions have sundered. Schütz's metaphysics of separate worlds deliberately—and fatally—disregards "the problems of the so-called sociology of knowledge."[36] The sociology of knowledge is no longer a "so-called" practice (if it ever was) but one that has become foundational to contemporary views of science in the world. In light of all that we understand of the spatial and social constructedness of science, it will no longer do to assert that treatment of a scientific problem is determined *solely* by the problem, the stock of pertinent disciplinary knowledge, and formalities of procedure. The conventions of scientific knowing always have a degree of kinship with those of the societies in which science is carried out. The question is no longer whether everyday activities can cross over into science, but only in what contexts and how they do.

In some cases practices of resident observing have been inadvertently begun when scientists pursuing incremental improvements in established practices found themselves doing something new and not quite orthodox. Small steps took them across an epistemic boundary they didn't know was there. Malinowski's discovery of resident observing is a good example. In the course of a standard ethnological survey, he began to do personal interviews and observing on site, to check on informants' veracity and to give a human face to the generic data of standard questionnaires. He didn't mean to stay for long in the Trobriand Islands, but the unexpected richness of the data that came from science in residence kept him there off and on for four years. In the "Here" of cohabitation, the empirical and theoretical value of participant observing became self-evident. William Whyte came in a similar way to his sociological version of participant observing. He went to Boston's North End to carry out a standard community survey, and to gain access to the community he began to hang out with a street-corner "gang" of young males, only to realize halfway into his project that these groups had become subjects of a resident sociology of small groups quite different from what he had planned. Recreations of everyday life serendipitously became his science.

In other cases investigators lacking formal training in a science became resident scientists directly through their life activities, thus bypassing the rules and restrictions of a discipline. Jane Goodall's route to resident science exemplifies this pattern. She knew nothing of academic ethology when she took up residence at Gombe and began to follow and watch chimps in their forest home. She did not mean to learn or improve upon an established science but wanted only to fulfill her youthful dream of living among wild animals in Africa. However, she was soon getting such a wealth of new and surprising information about chimp life that even doubting professional ethologists were won over.

Scientists' life activities also materially shaped their field practices. Jane Goodall is again an exemplary case. Her favored practice of habituation—openly and visibly following animals on their daily rounds and mimicking their social behaviors to get close—was one that professional ethologists generally shunned as inefficient and intrusive. Most likely it derived from Goodall's skill and experience in relating to companion species (horses, pets) and with the people in her unconventional life. Field practices of wildlife ecology likewise derived from practitioners' life experiences. Errington's fascination with the ecology of predation and his practice of "reading sign" came straight from his life as a commercial trapper and hunter-naturalist.[37] And Nels Anderson's uncommon ability to get life stories from itinerant hobo workers derived from his own life as a hobo and his habit of consciously—almost sociologically—entering into each new work culture as he encountered it on his winding life's path. Unorthodox life experiences and the relaxed rules of field situations afforded varied opportunities for individuals to evade disciplinary orthodoxies and turn activities of life into science.

RESIDING. We may not think of an activity as commonplace as residing as a scientific practice, but it can be—and not just a prologue to science, but science itself. The practices of residing are the means by which resident observers gather empirical evidence and form reasoned explanations of what they observe. Residing is relational: between observers and observed, and between both of them and their common neighbors. In the human sciences, resident observers work out a recognized identity, learn to respect local customs, and participate in appropriate ways in community life—much as their subjects do in life. As the sociologist Everett Hughes put it, learning to be a field observer is like learning to live in society.[38] I would add only that resident observing is not just *like* living in a society: it *is* living in a society. Maintaining effective relations with their subjects is how observers on the inside make residing into a scientific occupation.

The role of resident observer has been usefully assimilated to the sociologist Georg Simmel's concept of the social role of "stranger."[39] Simmel's "strangers" are not disconnected aliens just passing through. Rather, they constitute an organic element of communities, with a relation to native residents that is, paradoxically but usefully, both remote and close. Their freedom of mobility and detachment from communal commitments give strangers a unique ability to act in ways that settled members of a community cannot. Traders and moneylenders were historically the paradigmatic "strangers" in Simmel's sense, as were foreigners recruited to be judges or public officials, because they had no local pasts and loyalties to taint their neutrality. Resident observers can likewise be trusted by residents to observe without taking sides with any local faction. Observers' dual identity as neighbors and strangers from afar also enables them to give local communities a presence and standing in the wider world. Simmel's concept reminds us that the resident observer is not an anomaly but a scientific variant of a familiar social type: neither alien nor indigenous and both cosmopolitan and local—a bit strange perhaps, but no stranger.

The ethnographer is the prototypical resident observer—the type specimen of the species, so to speak. The villagers of Omarakana assigned Malinowski an identity and role—keeper of memories and stories—that was both familiar to them and fitting to what the ethnographer did every day: hanging out, gossiping, watching, asking questions, writing in his notebooks. Always present at dances and ceremonials, and at the unpredictable dramas of village life, he stayed largely on the sidelines and took no side. The ethnographer was part of the community, just one with an unusual identity and occupation. In a community where visitors from across the sea were routine—Trobrianders were renowned traders and blue-water navigators—a resident ethnographer, though a stranger, was not all that strange. The social practices of residing thus served the investigative purposes of ethnography. Modern societies have an even greater variety of people who come and go and who in their occupations exploit the advantages of being both settled and cosmopolitan, so that resident sociologists have more varied options of the stranger's role. They can hide, incognito, in some familiar local identity or, better, simply be themselves—academics working on degrees or writing books. That was how residents of Boston's North End understood and accepted William Whyte. Being openly and only himself—a stranger but not strange, without pretense—won the trust of his corner friends and subjects and made participant observing possible and fruitful.

But does this relational concept of residence apply when observers' subjects are free-living animals? Can observers be said to have a resident relation to creatures who normally flee at the sight of them? In some cases they can. Jane Goodall's practice of habituation was as much a relational practice of residing as anything anthropologists or sociologists did in their host communities. Physically Goodall and her coworkers lived more or less inside the chimps' forest home and practiced the conventions of chimp social relations to get close without provoking fight or flight. The practice of habituation marked her as a resident—a stranger but no longer alarmingly strange. To her subjects she became just another cohabiting forest animal: harmless and safely ignored. She had hoped at first for a relation of friendship but settled for toleration. Coresidence was possible even with creatures with whom habituation was not an option. With bobwhites a kind of coresidence was in fact unavoidable, because the species lives by preference in areas of small farms, drawn by the abundant food and cover. There scientific observers were coresident in a double sense: with bobwhites, whose hidden lives they sought to reveal; and with resident farmers, on whom observers depended for access, local know-how, and support. Stoddard and Errington were thus participants equally in a human and an animal ecology, and knowledge of both was vital to their science. As the English ecologist Charles Elton famously quipped, "When an ecologist says 'there goes a badger' he should include in his thoughts some definite idea of the animal's place in the community to which it belongs, just as if he had said 'there goes the vicar.'"[40] The thought extends to animal science the point that Everett Hughes made about sociology when he wrote that learning to be a field observer was like learning to live in society. Human and animal ecologies are not worlds apart.

OBSERVING. Observing is another activity that serves both everyday life and science. Unlike residing, it has been a subject of methodological reflection by practitioners since the seventeenth century and, recently, by historians.[41] The default view in modern science has been that observing is a necessary but lesser preliminary to the real science of experiment or hypothesis testing. This narrow role is the vestige of a grand and capacious domain that observers occupied in the age of Enlightenment—an "*empire* of observation," Lorraine Daston has called it. Observation then meant every kind of empirical work, including measurement and experiment, as well as inductive inference and hypothesis. Experiment was not yet the premier practice it would become but simply a specialized kind of observing peculiar to a few sciences like chemistry and pneumatics. Observation was everything that natural philosophers and natural historians did: in Daston's words, a "powerful, sophisticated, and deliberate" empirical practice and the "essential way of reasoning in the sciences."[42]

How this empire of observation devolved is a story that remains to be told, though in outline it is clear enough. Like the modern concepts of objectivity and experiment, the narrowed concept of observation was a product of the decades in which modern science was created out of natural philosophy and natural history, in the early to mid-nineteenth century. As experiment became a regular practice of many sciences, it came to be seen as best practice for all, leaving observation as an outmoded practice of sciences left behind by progress.[43] We see a snapshot of this devolution in John Herschel's influential methodological treatise of 1833. Herschel still defined experiment as a kind of observation, but now a superior kind that was "active" and akin to the calculated probing of legal cross-examination. Plain observation was in contrast "passive" and akin to a rambling and often obscure story, bits of which might one day by chance prove meaningful. Experiment made progress "rapid, sure, and steady"; in its absence, advance was "slow, uncertain, and irregular."[44] Thus was the epistemic foundation laid for a modern empire of experiment, and observation relegated to the byways of science and the ambiguous frontiers with everyday life. In extreme cases, like neoclassical economics, observing of actual economic behavior in life would be squeezed out altogether by mechanized polling and randomized statistical analysis. Observing in situ "lost its credentials" as a proper tool of science.[45] In the life sciences, "butterfly collector" was the stock putdown for those who persisted in observing.

Of course, practices of intensive and reasoning observation did survive. Observers of animal behavior in situ had never been passive watchers and recorders, one practitioner protested, but always active interpreters. In selecting what, where, and how to observe, they were as much active "participators" in their science as experimenters were in theirs.[46] Observing in this older style remained a mainstay of the human and social sciences, though always in the epistemic shadow of more prestigious metric modes. That, in Carlo Ginzburg's words, became the "awkward dilemma of the human sciences" (and, I would add, of resident sciences generally): "[S]hould they achieve significant results from a scientifically weak position, or should they put themselves in a strong scientific position but get meagre results?"[47] Ginzburg's encomium to the Sherlockian method of clues and conjectures articulates a concept of observing much like Daston's imperial mode, with its synergy of active observing and theorizing.

Resident sciences afford many instances of observing in that capacious original meaning of the term. Malinowski insisted his participant observing was a kind of empirical theorizing. Wildlife ecologists' forensic practice of "reading sign" was a scientific form of what hunters did in tracking prey—the ancient *fons et origo* of Ginzburg's method of clues and conjecture. In Goodall's practice of close-in, day-in/day-out walking and watching, new observations deepened previous conjectures, and conjectures sharpened her eye for significant occurrences. It was the same with William Whyte's thoughtful observing of his corner cliques. Resident observing is in general a practice of reasoning not from abstractions but with particulars—a practice of clues and conjectures. It would not have seemed strange to citizens of that earlier and grander polity of knowledge.

STORIES. This book is full of stories: observers' stories of their subjects' activities and their own as observers and neighbors, and of the life paths that brought observers to places of coresidence. We read Malinowski's eyewitness stories of Trobriand village life, Nels Anderson's life histories of sixty particular men who became and remained hobos, William Whyte's stories of his activities and those of his corner "boys," Jane Goodall's village tales of her chimps' distinctive personalities and lives, stories of how Herbert Stoddard's and Paul Errington's activities as hunter-naturalists became their ecological science. Stories are not just a way of presenting known facts and concepts: they are also the means by which facts and concepts are produced. Observers explore and explain history and science by telling stories.

A premise of these case studies is that narratives are a true and proper means of understanding human and natural worlds: they may be faulted on facts or logic, but they cannot be dismissed as untrue just because of their narrative form. For most of the long time in which histories and natural histories have been written, that was the accepted view. But from time to time, and strongly in the middle decades of the twentieth century, the propriety of narrative in history and in science has been vigorously contested. The defense, it is safe to say, has prevailed among professional historians and is gaining in science as well. But a brief recounting of the issues is in order.

Arguments for and against the fidelity of narrative history to the actual world have generally turned on the fact that narrative form—causes and consequences; beginnings, middles, and ends—is the familiar mode of imaginative fiction and everyday social relations. That conformity with common experience was once what gave narrative historians their authority. "[N]arrative explanation is satisfying precisely because it never strays far from ordinary discourse," the philosopher David Carr writes. It places actions "within a familiar repertoire of actions, emotions, and motives. These are things we've seen before, and we illuminate the unfamiliar by relating it to the familiar." "Familiarity is reassuring, especially when contrasted with the prospect of veering into the hidden and the arcane."[48] Natural historians' stories of the natural world persuade for the same reason: though animals may lack conscious motive and calculation, we understand their activities in much the way that we understand our own, as unfolding actions and events. Stories—literary, historical, social, scientific—were once the shared mode of thought of a common culture.

That logic of credibility was upended in the twentieth century as humanities and sciences were organized in distinct professional institutions. In the modern view, history and science were superior ways of knowing because they were skeptical of what was taken for granted on the basis of individuals' experience. Historians and scientists sought warrants of truth not in common experience and common sense but in their own communal procedures of disciplined empirical study and reasoning: not common, but uncommon sense. Narrative was sidelined as artistic creation and an ex post facto imposition on phenomena that have no such tidy structure and meaning. Narrative histories were in their very form a distortion of worldly realities and *in principle* could not be true to the events they purported to explain.[49] Narratives are not lived but told; they are art, not knowledge. One needn't know the particulars of stories to doubt their truth: their form alone was evidence enough.

This banishment of narrative from history rests on the same assumption as Alfred Schütz's banishment of everyday practices from science: namely that disciplined knowledge making and everyday life are separate and incommensurable worlds. Everyday occupations and activities cannot be turned to science, because by definition there is no place for personal matters in scientists' working "Here." Storytelling can likewise have no place in history, because it is a practice of artistic imagining: true to life, perhaps, as in a novel; but not true. These versions of "separate worlds" have the same fundamental flaw: they are belied by the reality of practice. It may in theory be impossible for scientists to use everyday practices, or for narrative historians to depict truly how human history works—yet they do it all the time.[50] And they are able to do so because history (like science) is not a world of its own, separate from the world of everyday life.

As David Carr has decisively argued, narrative form is in fact derived from worldly actions and events, not imposed on them in retrospect. Nature and society as we perceive them have an inherently narrative form.[51] To briefly summarize Carr's argument: In our lives we are always in the middle of some action or event, with a beginning, middle, and end. This narrative patterning is not created after the fact by those who write about it, but is created in the process of living in the world. Both as individuals and as members of groups, we experience and understand actions and events in narrative form. "[W]e are constantly striving, with more or less success, to occupy the story-teller's position with respect to our own lives." When we are asked to explain our own or others' actions, Carr notes, the expected answer is typically a story. We tell ourselves stories to become clear on what we are doing, so that "narrative activity is a constitutive part of action, and not just an embellishment, commentary, or other incidental accompaniment." Narrative activity is thus "practical before it is cognitive or aesthetic; it renders concerted action possible and also works toward the self-preservation of the subject which acts." Narrative is not a description of things that already exist but takes shape in the making of those things in the first place. It is, in other words, "a mode of being before it is a mode of knowing."[52] The structure of narrative is the structure of actions in the world, and that is why "we can genuinely be said to explain an action by telling a story about it." Narrative is how the world exists to us and how we operate in it. So it is not narrative that needs to be justified, but rather its absence: "[I]f we depart from a common-sense mode of explanation, such as narrative explanation, in favor of another model, we had better have good reasons for doing so." Stories and actions are not separate worlds: it is "narrative all the way down."[53]

Carr's defense of narrative history is easily extended to the human and social sciences. But does it extend as well to the science of species that do not (so far as we know) live stories, or to sciences of inanimate things? As Mary Morgan and Norton Wise have noted, historians and philosophers of science have with scattered exceptions paid little attention to narrative as a practice of science.[54] The last few years, however, have seen a strong and growing revival of interest: in conferences, special sessions of society meetings and journal issues, and programs of systematic study.[55] Morgan and Wise became aware of narrative elements in seemingly unlikely subjects. Morgan noticed that economists generate stories from mathematical models to explore theoretical potentials and limitations, and that they insert particular stories into theoretical models to test their fit with actual economic phenomena. Wise likewise perceived a narrative structure in theoretical physicists' simulations using mathematical models to represent processes too complex for the reductive laws of classical or quantum mechanics.[56] And if we find narrative in these most leanly abstract of sciences, must we not expect to find it as well in the many sciences that are richly empirical and situating? As Wise notes, "[N]atural scientists today are much more likely to appeal to models to ground their claims to understanding and explanation than they are to appeal to natural laws."

And along with models comes a narrative mode of understanding natural phenomena.[57] Morgan's roster of sciences with narrative elements covers a very broad swath indeed.[58] Narrative in science is not exceptional, it seems, but the rule.

Morgan and Wise make their case for narrative in science in the same way that Carr made his for narrative history, reasoning that a narrative form is inherent in the structure and operating principles of nature, as it is in human affairs. Physical and biological processes are almost by definition developmental and historical in form, with causes and effects and beginnings, middles, and ends. Narrative accounts, here again, are not cultural impositions on phenomena but derive empirically from them. For complex phenomena especially, narrative affords open and flexible ways of ordering tangled banks of facts that reductive laws and theories cannot. "[A]ll too often theories are too thin to cover the problem or the ground," Morgan and Wise write. "Narrative then provides a natural form for bringing related elements into order or creating order out of disordered materials."[59] It is the orderly disorder of nature itself that gives scientific stories their authority and explanatory power.

Narrative is thus not an imperfect stand-in for hypothetico-deductive analysis but in many cases the preferred route to understanding nature. It is deduction from simplifying laws that is the shortcut, and it is taken at the cost of leaving out the very elements of context and history that are vital to real understanding.[60] Abstract deductive science is parasitic on those sciences that observe and tell stories, not the other way around. Wise thus upends the conventional modern order of standing among the sciences, much as Samuel Gorovitz and Alasdair MacIntyre did in arguing for a science of particulars. And as for resident science—preeminently a science of particulars—narrative is not just one way, but possibly the best way, of understanding what is observed in the particular situations of the world. So perhaps the cases of resident science in this book are not outliers of a modern empire of monocultural abstraction, but growing points of a polycentric and polyglot empire of observing and conjecturing in place.

CHAPTER 2

PARTICIPANT OBSERVER

BRONISLAW MALINOWSKI

When friends ask what I mean by "resident" observing, I usually point first to Bronislaw Malinowski. They may not know much about the man, but he remains to this day an iconic anthropologist, one of those who live among strange peoples to understand their ways of life. His five monographs on the Trobriand Islanders—on spirits of the dead and ideas of reproduction, the kula trade cycle and blue-water navigation, sex and family life, customary law and transgression, and gardening and provisioning—are engaging works of science and literary art. And the man himself was an intriguing work of art. A polyglot blend of national cultures (Polish, German, English) and sciences (philosophy, physics, anthropology), he was charismatic and exasperating, forthright and duplicitous, generous with students but aggressively combative with rivals. He inspired intense loyalty in some and in others a visceral enmity, and was always a focus of controversy. In his prime he was widely regarded as the inventor of the discipline-defining practice of participant observing—the type specimen, so to speak, of the genus Ethnographer. Yet in his later years and in the decades following his death, his place in the history of social anthropology has been vigorously challenged.[1]

That controversy centered on Malinowski's efforts, as a professor of social anthropology at the London School of Economics, to create a general "functionalist" theory of culture, and on the resulting contests with his archrival, A. R. Radcliffe-Brown, whose functionalism was more formally theoretical. These disputes in effect engendered two Malinowskis: the brilliant and imaginative empirical field-worker and the forgettable theorist. As his student and colleague Edmund Leach wrote, "For me, Malinowski talking about the Trobrianders is a stimulating genius; but Malinowski discoursing on Culture in general is often a platitudinous bore." Malinowski's functionalism was to most practitioners "repugnant," Leach went on, and his abstract theoretical writings "are not merely dated, they are dead."[2] Yet his Trobriand monographs are to this day still read and admired for their vividly concrete accounts of Trobriand village life and their lasting insights into human social behavior, even as his biocultural functionalism languishes in deep academic disfavor.

This schizoid picture of two Malinowskis—the empiricist and the theorist—is in important respects a misleading one, especially in its dichotomous view of descriptive and interpretive science. As several commentators have pointed out, Malinowski the empiricist *did* theorize: not with the abstractions of social theory, however, but empirically from observed particulars. The concepts that animate and give general meaning to his Trobriand monographs derive not from theorists like Émile Durkheim (Radcliffe-Brown's favorite thinker), but from his resident observing. In Carlo Ginzburg's terms, he operated in the Sherlockian mode, from clues to conjectures. Malinowski's later fling with a more abstract "Galilean" mode of science has diverted attention from his distinctive practice of empirical reasoning. The discovery of that combined practice is what made Malinowski the iconic Ethnographer. And that discovery is my subject here. How did it come about?

Let us begin on the beach at the Trobriand village of Omarakana, with Malinowski's famous evocation in *Argonauts of the Western Pacific* of his maiden experience of fieldwork:

> Imagine yourself suddenly set down surrounded by all your gear, alone on a tropical beach close to a native village, while the launch or dinghy which has brought you sails away out of sight. Since you take up your abode in the compound of some neighboring white man, trader or missionary, you have nothing to do, but to start at once on your ethnographic work. Imagine further that you are a beginner, without previous experience, with nothing to guide you and no one to help you. . . . I well remember the long visits I paid to the villages during the first weeks; the feeling of hopelessness and despair after many obstinate but futile attempts had entirely failed to bring me into real touch with the natives, or supply me with any material. I had periods of despondency, when I buried myself in the reading of novels, as a man might take to drink in a fit of tropical depression and boredom.[3]

He further invites readers to imagine his initial struggles in pidgin English to record native customs and culture but getting only native boilerplate and jumbles of facts, in which he could find no meaningful order. His European hosts he found pleasant enough, but their cultural aims and biases made them poor guides to native life. It was not until he was by himself in the field, he tells us, that he began to learn the "ethnographer's magic"—that is, the firsthand field craft that brought to life "the real spirit of the natives, [and] the true picture of tribal life." And the chief principle of that craft, he writes, was residing in a native village, where he could get to know his neighbors, learn their language, converse, observe, and be part of their everyday lives.[4] The "magic" began with being inside his object of study and getting the inside stories.

The scenes that Malinowski evokes in this storybook opening of the book that made him famous are in fact not those of his arrival in the Trobriands, but rather of his initial, hopscotch survey of villages along the coast of eastern New Guinea. It was later, in residence in Omarakana, that he experienced resident observing and discovered its scientific "magic." What we have in the famous introit to *Argonauts* is a collapsed and schematic account of the whole experience of becoming an ethnographer, from uncertain apprentice to confident master. Malinowski's account has been read by some as a rhetorical construction of a self-made hero, but that seems to me not quite right. The figure he presents seems less a hero in embryo than an ill-prepared and self-doubting beginner. And the purpose of his little confession seems less self-fashioning than pedagogical: to advise would-be ethnographers that the craft is learned only inside, by trial and error; and that this practical apprenticeship does in the end work, so don't give up.

The persona of the Ethnographer was a creature of its time, and as times changed so did anthropologists'—and historians'—readings of it. In the current revised view, Malinowski was not the sole inventor of resident ethnography but one of a community of mainly English ethnographers who were all moving more or less in that direction. The Ethnographer was in this view a founding myth; and Malinowski, a mythmaker who achieved the iconic status of founder for the contingent reasons of a vivid and visible personality, accidents of social location and timing, and skill in self-fashioning. He was advantageously placed in his profession, had students to spread the word, and knew how to tend his reputation and career in a contentious academic culture. Participant observing was in this revisionist view not "discovered," but evolved gradually through the work of many hands into a settled and productive methodology.[5] Malinowski was thus more midwife than heroic progenitor—a man at the right place at the right time to stand in for all.[6] And he aided in the later deconstructing of his mythic standing by, as one second-generation ethnographer put it, his "histrionic, not to say exhibitionistic" behavior, insisting on the revolutionary novelty of his work and denigrating his competitors.[7]

Demythologizing heroic founder figures has been a useful move toward a contextualized history of science, and so it is here as well. However, turning discoverers into self-fashioning mythmakers may also obscure vital aspects of historical reality. Creative and lucky individuals *do* transform their sciences, and concepts of discovery or invention usefully remind us that science is carried on by people who have agency and change their worlds, and often have that agency by virtue of being iconic personages. Raymond Firth has argued that Malinowski's dramatization of his field experience was a vivid literary representation of the ideals and practices that shaped generations of field-workers.[8] It was, in other words, meant to epitomize and instruct. The myth was meant to do real work, and did. We should also bear in mind that deconstructing "myths" is a favorite means of historians' own professional self-fashioning and is itself open to skeptical deconstructing. It is perhaps time for a historiographical rebalancing.

In this account I will view Malinowski in the context of a community's evolving field practice, with the aim of understanding how in the Trobriands in 1915–18 he made a decisive turn in that practice.[9] Such events occur at conjunctures of an evolving science with individuals' unfolding life histories. The task for historians is to follow the paths of both into situations in which change is unexpectedly brought about. If Malinowski had not been at the right spot at the right moment to see what extended residence could do for ethnography, someone else would have been. But as it happened, the person in that situation was an intense, driven, neurotic, brilliant, ambitious, thin-skinned, hypereducated, Anglified, Polish ex-philosopher humanist and artist manqué.

Resident observing, I will argue, grew out of an improved, intensified variant of traditional ethnological survey, which had some of the elements of the later practice but not the crucial one of residence. "Intensive local investigation" was Malinowski's own term for what he did ("participant observation" was a later usage), and it better captures the liminal situation in which an established activity morphed into something new and different.[10] To understand how intensive survey became resident and participant ethnography, we must look into the practices that were state of the art at the time of Malinowski's arrival in the Trobriands, and the situation in which a transition from peripatetic survey to residence was a surprisingly small and easy step—almost a kind of sleepwalking.

ETHNOLOGICAL SURVEY TO ETHNOGRAPHY

If one trend marked the history of ethnography from the 1870s to World War I, it was ethnographers' persistent efforts to make methods of empirical fieldwork more systematic and exact. The history of these early decades of social anthropology has been well and fully told by historians of anthropology George Stocking, Henrika Kuklick, and others, and my brief synopsis here rests on their work.[11] The first ethnologists were armchair writers and social philosophers who did not go afield themselves but harvested facts from the reports of explorers, travelers, and others who did. In that context empirical improvement meant exercising some control over the haphazard flow and uneven quality of secondhand information. One such improvement was to write and distribute handbooks to instruct travelers and residents in far-flung places in the proper methods of questioning native informants and in established categories and topics of particular ethnological interest. *Notes and Queries for the Use of Travellers and Residents in Uncivilized Lands* (1874), renamed (pointedly) *Notes and Queries on Anthropology* in the second edition of 1892, was the prototype of this genre.[12] How all this good advice was actually used was of course beyond ethnologists' control.

Improving ethnologists also became more discriminating in their sources. They relied less on accounts of travelers and adventurers—notorious embellishers of truth and tellers of tall tales—and more on long-term European residents—missionaries, traders, colonial administrators. They were more likely to know native languages and to have regular contacts with native residents, so were better able to identify trustworthy informants and witness their ceremonies firsthand. Such transcultural mediators were everywhere in the age of empire; and some of them, missionaries especially, produced valuable ethnographic work.[13] Malinowski later gave these "old recorders" credit for seeing and reporting, in their day-to-day dealings with subject peoples, the "seamy" or lived side of their life, even if they might not perceive the sociological order in the apparent chaos. He credited one of his own contacts, the Reverend Mathew Gilmour, with knowing more about Trobrianders than anyone else, including himself.[14]

There were limits to this kind of ex post facto improvement, however. Resident Europeans' perceptions of indigenous peoples and their activities were inevitably shaped by their own culture and professional duties of regulating and reshaping the behavior of subject peoples in matters of economy, religion, and law and order.[15] The social categories that Europeans took as universally human—patrilineal inheritance, patristic theology, legalistic land title, a narrowly utilitarian conception of trade—blocked imaginative entry into the worldviews of societies in which such alien categories were literally unthinkable. Malinowski met resident traders who spoke Trobriand better than they spoke English and could comport themselves appropriately in ceremonies, yet were hopeless as ethnographic informants, because they could not enter into natives' conceptions of their social and cosmic worlds.[16] Residence did not in itself make a man a resident ethnographer or even an apt partner for ethnographers.

A more far-reaching, proactive improvement in empirical practice was carried out in the 1890s and 1900s, when ethnologists themselves began to go afield to gather data and see for themselves, either as individual travelers or in organized expeditions, of which the famous expeditions to the Torres Straits in 1889 and 1898 became the model. Ethnologists who till then had been secondhand consumers and interpreters of facts now became firsthand producers as well—that is, ethno*graphers*.[17] This consolidation into a single role of what had been two separate ones was not unique to anthropology but was a trend in all the sciences that collected facts or things in the wide world. For example, museum curators and taxonomists were another group who acquired the right and the means to leave their offices and collect in situ.[18] The advantages of a situated role were obvious: scientists could choose and control their sources and methods, inject scientific purpose and system into what for others had been sideline activities, and adapt preset agendas to seize opportunities that turned up unexpectedly in the work.

The prevailing mode of empirical inquiry in pre–World War I ethnology was ethnological survey. This package of theory and empirical techniques was perambulatory and wide-ranging over entire cultural areas, rather than intensive examination of particular locales. The chief technique of ethnological survey was interview, usually of trained and paid informants. This was not a situated practice. Interviews were typically carried out not on informants' home ground but on a local trader's or missionary's veranda or in ethnographers' camps, and they followed the European categories of *Notes and Queries*, not informants' own. Survey ethnologists did on occasion visit native villages, especially after 1900, but visits were generally brief and supplemental to interview, to check informants' veracity and watch native ceremonies. Survey ethnology was not a science of observing and partaking. It was a mobile, expeditionary science of collecting mainly verbal records complemented by material artifacts, photographs, anatomical and psychological measurements, and so on: naked data without full contexts of use.

Survey ethnology was also a historical science. Its aim was to use the evidence of contemporary "primitive" societies to reconstruct the evolution of human society in prehistory—an odd mixture of Darwin and Genesis. Although this historic*ist* belief in a universal scheme of social evolution was eroding by the turn of the century, the aim of ethnological fieldwork remained historic*al* into the twentieth century. If native societies could no longer plausibly be regarded as survivors of the early stages of a pan-human social evolution, they could yield evidence of actual histories of human migrations, displacements, and cultural and artistic mingling like those that had occurred on a vast scale in the not-too-distant past in the South Seas archipelagoes. This actual history, unlike the imagined evolutionary one, was empirically based; conjectures could be framed and tested by ethnographic investigation.

Then, in the 1920s and 1930s, historical ethnology was in turn superseded, or overlaid, by a sociological ethnology that sought to understand how human communities variously operated in the here and now. A diachronic science gave way to a synchronic one. The intellectual history of this transformation has been well told by George Stocking, Henrika Kuklick, and others; and there is no need to recapitulate it here.[19] The point is simply that in both systems conceptual aims and empirical practices were symbiotic, the one justifying and sustaining the other. Thus historical reconstruction required extensive knowledge of locales in whole culture areas; and extensive survey using *Notes and Queries* interviews produced data efficiently and in categories ready-made for comparative analysis. A sociological anthropology, in contrast, required a deep knowledge of local particulars, which could be acquired only by sustained firsthand study in a few places. The richly particularistic data produced by resident study impeded historical ethnology, but it made a sociological ethnography possible.

The question then is, What enabled an existing self-sustaining system of ends and means to become a new one? It was those occasional in situ interviews of survey, I think, that were the bridge. A minor practice of survey ethnology became, unexpectedly, the major practice of resident ethnography. Incremental improvement in survey data gathering had by the early 1910s reached what in hindsight was a tipping point, where in the right circumstances a further small change could turn survey into resident science. It is useful in visualizing this transition to recognize an intermediate category of field practice—*intensive survey*—that was still historical in aim and extensive in practice but that fostered local situations in which visiting ethnographers might experience firsthand the unsuspected advantages of resident observing. Intensive survey improved the odds that the hands-off formalities of *Notes and Queries* interview would slide spontaneously into the personal informalities of neighborly conversation and participant observing.

This transition did not develop out of thin air. Ethnographers had been thinking about intensive local study practically since the first Torres Straits expedition of 1889; but actual residence by ethnographers always seemed impracticable. In the 1892 edition of *Notes and Queries*, for example, Charles Read observed that to get even superficial information on the many aspects of a native community would require an impossibly long and devoted residence in the place.[20] He may have had in mind the expat way of life of resident traders and missionaries. Some anthropologists looked not to extended residence but to an intensified expeditionary mode. In 1905 and in 1906, Alfred C. Haddon called for "intensive studies of restricted areas" using systematic recording of genealogies, asserting that it was "only by careful regional study that the real meaning of institutions and their metamorphoses can be understood."[21] His words sound prophetic of resident observing, but what he had in mind was an extended—and prohibitively expensive—expeditionary survey. Such a project, he noted, would require a ship designed and fitted out for survey science, with a permanent scientific staff who would pursue a long-term campaign of regional data collecting. Haddon noted as well that intensive study would be possible only in locales that did not require inordinate outlays of time and energy, given the many locales that would be surveyed.[22] In short, he envisioned a state-sponsored expedition like those to the Torres Straits but on a far larger scale—a dream he must have known was unlikely to be realized any time soon. He and his colleagues W. H. R. Rivers and Charles Seligman did manage to send individuals afield for extended periods, including Gunnar Landtman, to Kiwai island (1910–12); but Landtman's task was, typically, to fill in gaps in previous expedition surveys.[23] No one at the time foresaw the one-man or one-woman resident ethnography that would unexpectedly become the new default practice of ethnographic science.

What Haddon's wishful thinking may have done, however, was to give intensive local ethnography a categorical reality distinct from existing practice. So in 1912, when Rivers renewed the agitation for intensive local investigation, he could give it a name and an identity distinct from traditional survey. "[E]thnological inquiry has two main varieties," he wrote, "which may be called 'survey work' and 'intensive work,' respectively." Whereas survey ethnographers covered a wide territory and favored some special aspect of community life (language, religion, kinship, material culture), intensive field-workers would be generalists, who would live for a year or more in a native community of four to five hundred people, engage personally with their subjects, and document every aspect of their lives. Yet for Rivers, as for all other ethnologists (including, at first, Malinowski), the purpose of intensive local study was still to reveal gaps and mistakes in the accumulated data of expedition surveys, and to keep ethnologists in mind of how much remained for them to record and inventory.[24]

Indeed, large-scale survey seemed ever more urgent in the prewar years, as settler colonies overran indigenous cultures that were then gone forever. If ethnology was to have a future at all as an empirical science, it would have to be a science of extensive survey and efficient data collecting. Rivers reckoned in 1913 that there was about a thirty-year window of opportunity between the time when "savage" societies were too wild for ethnographers to visit and the time when they were spoiled for science by culture contact. Melanesia was then just in that window, he thought, whereas interior New Guinea to the west was still too untamed and dangerous, and Polynesia to the east was already too Europeanized.[25] Intensive survey, not deep local study, was in this situation the obvious method of choice.

However, practices of intensive survey were moving toward a resident ethnography. One such practice was Rivers's famous "genealogical method," in which interviewers systematically collected data on kinship relations, not from a few selected informants but from as many as possible, cross-checking and correcting errors and omissions until ethnographers understood a community's kinship structure as fully and intimately as its members did. Though positivistic in spirit and technique, the genealogical method did bring data collectors closer to their subjects' experience and views. Natives respected white strangers who knew more about their genealogies than they did themselves, and ethnographers began to think in natives' social categories rather than their own. As Rivers noted approvingly, they were entering into Melanesians' own obsessive concern with genealogy, "using the very instrument which the people themselves use in dealing with their social problems." Total knowledge of kinship relations would enable ethnographers to perceive the unspoken social meanings of ritual ceremonies and even "formulate laws regulating the lives of people which they have probably never formulated themselves." For example, ethnologists untangled marriage laws so byzantine that not even long-term European residents had been able to make sense of them.[26]

Improvements in the techniques of formal interview also brought survey ethnologists closer to their subjects' view of their world: conducting them in native languages rather than in pidgin English or via interpreters, following informants' leads rather than hewing to a preset script, and framing questions around particular occurrences rather than in abstractions, which for natives were an alien way of thinking and typically evoked baffled silence. With concrete particulars, informants could engage as naturally and spontaneously as they did in village daily life. As Malinowski put it, "A real case . . . will start the natives on a wave of discussion, evoke expressions of indignation, show them taking sides," thus revealing unpremeditated information on the variability of beliefs and on behind-the-scenes causes of social actions and events. Interviews could thus become a novel—yet familiar—activity of villagers' daily lives, not just the extramural occupation of a few trained and paid informants.[27] In 1912 the armchair Oxford ethnologist Robert Marett anticipated that observing community life up close would show that native customs were not as rigid and unchanging as informants claimed, but were continually modified as individuals adapted them to their own particular ambitions and desires.[28] It was a prescient harbinger of what Malinowski's resident study would in fact reveal.

Yet a fully resident practice remained beyond prewar ethnographers' imagining, even as their practices edged toward it. Marett advocated intimate study of daily life as "supplementary work": meaning, presumably, supplementary to regional survey.[29] Rivers likewise meant his genealogical method to make survey more efficient and precise by freeing ethnographers from the necessity of learning native languages and laboriously cross-checking informants' testimony. Intensive local study was meant to improve survey practice, not replace it.[30] However, it created situations in which visiting ethnographers could relate to their subjects in a way that was close to that of a resident observer. In such situations lengthy residence could be not just an unattainable ideal but a real option; thresholds to change were low. Malinowski's winding path to anthropology intersected with that science at just that moment and in just that situation.

MALINOWSKI: LIFE HISTORY

Though Malinowski discovered participant observing by doing, the discovery owed a good deal to the intellectual baggage that he brought to the field from his own considerable life experience (he was then thirty years old). He was diversely educated in physics, mathematics, and philosophy of science, as well as humanities and the human sciences; and had a sophisticated conception of scientific method. Trained and experienced in critical analysis of scholarly texts, he had authored some dozen publications of his own—reviews, articles, books—in three languages (he knew six in all). He was a cosmopolitan in the continental mode, with an old-world etiquette (less "bourgeois" in manner than "ancien régime," his daughter wryly observed)—a citizen of the prewar pan-European intelligentsia.[31]

He brought to his newly adopted science two methodological commitments from his academic experience. One was a conception of anthropology as a sociological science. As Robert Thornton firmly put it, "Malinowski's advocacy of 'holism' and 'function' . . . was based on philosophical commitments that had been made well before his being set down on a tropical island. . . . [H]is experience in the field was not the cradle of functionalism, as he himself so frequently asserted."[32] He brought as well a reasoned and vehement antipathy to the older view of ethnology as a historical and evolutionary science. The purpose of anthropology in his view was to work out the structures and operating principles of human communities in the here and now. Unlike his mentors, Malinowski never experienced a transition from a diachronic to a synchronic view: he was already a synchronist when he began his apprenticeship as a producer of facts in the field.

He also brought with him a deep commitment to empirical science and rigorous empirical methodology. Mixing factual work with speculation was, as he later put it, "an unpardonable sin against ethnographic method."[33] He thus shared the empiricist temper of English ethnologists trained in natural science, like Rivers (psychology) and Haddon (zoology), though his empiricism had its roots not in the science of lab and clinic but in philosophy of science, and especially in the positive philosophy of the physicist-philosopher Ernst Mach. Malinowski's empiricism was, however, less puritanically positivistic than either Rivers's or Mach's and more humanistic and literary.

What Malinowski did not have before New Guinea was actual experience of gathering original facts in the field. He was in 1914 a library empiricist: a skilled consumer and user of other people's facts. In producing facts he was untrained and untried. He did, however, have experience of failing to produce original facts—in a lab, in physics. And a nagging uncertainty about his capacity for empirical production persisted into his first months as an ethnographer in the field. That failure is crucial to understanding how Malinowski became an ethnographer and of what sort.

Malinowski grew up in Cracow, the intellectual and artistic center of the Polish province of Galicia in the Austro-Hungarian Empire. His father was a professor of Slavonic philology and folklore at the Jagiellonian University, and young Bronio's early education was the classical regimen of a youth intended for an academic career: Latin and Greek (of course), Polish and German (the family spoke French at home), and mathematics and physical science. Summers he spent in Zakopane, a flourishing artist colony in the Tatra Mountains, where his closest friends were artists and bohemians. A particular friend was Stanisław Witkiewicz—Staś—a precociously talented boy who grew up to be a famously imaginative and original painter, while Bronio took the opposite course, into natural science, though without ever losing his taste for literary art. (He had tried without success to produce works of art and literature as good as those of Staś and his other artistic friends at Zakopanie.)[34] Staś was in a way the other half of Malinowski's scientific-artistic self, the two sides always a bit at odds, like sibling rivals. The tension in Malinowski's science between rigorous empiricism and imagination thus had deep emotional resonances of personal identity and self-worth.[35]

At the Jagiellonian University in Cracow (1902–6) Malinowski devoted himself to his chosen specialty of mathematical physics and philosophy of science. (His courses were one-third philosophy, one-third physics, one-quarter mathematics, and one-twelfth everything else.) The science was abstract and theoretical—thermodynamics, electrodynamics, theoretical mechanics—and so was the philosophy of science: a rigorous neo-Comtean positivism, with Ernst Mach as reigning authority and focus of philosophical disputation. Malinowski's view of scientific method was formed in that student encounter with Mach.[36] Mach believed that appeals to "nature" were no warrant for the truth of scientific propositions—because humans do not know nature directly but only their own sense impressions and logic of reasoning. Thus the truth of knowledge claims was measured solely by the intellectual qualities of the claimants and their skill in empirical perception and inference, because these and these alone could be assessed objectively as observable facts. Direct appeals to nature were not science but "metaphysics." Malinowski rejected Mach's doctrinaire positiv*ism*, which notoriously denied natural reality to all human conceptions (e.g., atoms). Imagination was essential for science, Malinowski believed, so long as it was disciplined by Mach's critical empiricist methodology.[37]

This reliance on critical method afforded novices a relatively straightforward entry into the community of philosophers: one read and mastered the canonical literature of a subject, seeking and then correcting some lapse in facts or reasoning. Finding a flaw in a master's thought established a novice's credibility as a new member of the community of philosophers: it was the ante into their communal game of knowledge making. Malinowski demonstrated his bona fides in a dissertation on Mach and the Machian Richard Avenarius, where he exposed a vestige of "metaphysics" in Mach's appeal to the physiology of sense perception as a warrant for the truth of sensations—a forbidden appeal to biological nature.[38] This ingenious corrective was Malinowski's ante in. He produced no new material facts about physics, and at that stage in his apprenticeship he did not need to. One took the first steps to being a producing philosopher-physicist not in a lab, with instruments, but in a library, with texts. The lab came later.

From the Jagiellonian Malinowski proceeded to the University of Leipzig for advanced study in thermodynamics, the subject he had chosen for the *Habilitationsschrift* that would open the door to an academic career. That was what his mentors expected, and for him it was, as he put it, the path of least resistance.[39] However, experimental physics proved more difficult to enter than philosophy. He disliked Leipzig, for both personal and professional reasons, and though he had signed up for three semesters of study, he left (or was dropped for lack of attendance) after only two, abandoning physics and Germany for England and human science. There is surprisingly little direct evidence of how and why Malinowski turned to social anthropology. What he said about it was cryptic and evasive: that his "intellectual dilettantism" had tipped him over to social science; that his youthful infatuation with the easy refinement and social grace of the English had drawn him to London.[40] These seem hardly reason enough for such a drastic change of life. Historians have sought a more definite attraction in the ethnopsychology and ethnoeconomics of Wilhelm Wundt and Karl Bücher, whose lectures Malinowski heard at Leipzig. But his biographer Michael Young argues that these aging scholars were too devoted to evolutionary historicism and too uncritical in their use of secondhand facts to be attractive role models. Young suggests that Malinowski just gradually stumbled into anthropology, and I think that is close to the truth.[41] However, there may have been a particular push and pull that gave his life's path an unexpected turn.

Malinowski learned in Leipzig that he disliked laboratory work and had no aptitude for it. He could not make the required career transition from user to producer of original facts. What in philosophy he could do with relative ease in a library with texts, he could not do in a lab, with machines. Clues in diaries, letters, and fragments of autobiography attest to his sense of a door closing in front of him. He constantly enjoined himself to do better in the lab: to concentrate, work more calmly and steadily, abstain from diverting chitchat and friendships, avoid daydreaming and mental slovenliness, and stop trying to sham and show off. Toward the end of his first year in Leipzig, in November 1909, he dejectedly reported that his work in the lab was poor; he could not remember what he had done there but thought he had probably just wasted time.[42] He was, in short, not producing the empirical stuff required for a *Habilitationsschrift*. He had no future in empirical natural science: it was for him a dead end.

Malinowski had at the time been reading somewhat aimlessly in various subjects (the "dilettantism" for which he faulted himself then but later credited for turning him to anthropology), including the ethnosocial sciences of Wundt and Bücher. In addition to their intellectual interest, human sciences had the practical advantage of being more accessible to novices than physics, and in a way that was familiar to him. Like philosophy of science, they could be entered in the library by methods of textual analysis, in which he had already proved his expertise. Human science had the additional advantage of combining Malinowski's contending interests in science and literary arts. As he confided to his diary while struggling in the lab, "I wish to collect [scientific] material. I wish to go on acquiring a capacity for [literary] expression. A complete loss of mastery of the word would be simply fatal for me."[43] Anthropology thus became the path of least resistance—indeed, the only path—to habilitation and an academic career, replacing heat physics. The one was blocked; the other, wide open. Failure in the lab was the push to a drastic change of life course; the prospect of success in the library was the pull.

This is conjecture, to be sure; but there is evidence for it in the written works that mark Malinowski's path from philosophy and physics to anthropology, especially his first published book in ethnology, *The Family among the Australian Aborigines* (1913). As Michael Young tells us, this work grew out of a dissertation in *Völkerpsychologie* that Malinowski began in Leipzig in 1909 and took to London in 1910, where it "transmogrified" into a critical commentary on the literature of a core subject of London-Cambridge anthropologists, in their signature style.[44] The book is no less exemplary of the style of critical commentary that Malinowski had made his own as an apprentice philosopher: a comprehensive study of a scholarly literature, with critical analysis of authors' facts and logic, and an exposé of any lapses in fact and reasoning. It was the method of his thesis on Mach applied to ethnology, displaying the novice's mastery of his new field of study and his ability to contribute to the common enterprise. It was in other words his ante into the community of ethnologists. *The Family among the Australian Aborigines* and other critical essays from his London years took Malinowski with relative ease from philosophy to anthropology. No work in either lab or field was required, just the humanistic method of textual analysis applied to texts in the British Museum Library. His unhappy encounter with laboratory physics thus proved to be a brief, though formative, detour in his life's path.

Malinowski's reworking of data on the Aboriginal family was in one significant way an advance on his critique of Mach: it operated not on abstract propositions but on material facts. In winnowing sound from unsound field data he was not just consuming facts but creating them. Malinowski was aware that what he was doing in the British Museum Library went beyond what he had done as a philosopher: it was "a new way of working in the library," he noted, that involved "independent research with the concrete idea of publication."[45] Indeed, his book on the Aboriginal family can be understood as a kind of virtual fieldwork: a library dress rehearsal with secondhand facts for the firsthand production of facts in the field.

Malinowski the anthropologist never gave up the humanistic practice that gave him entry into anthropology. In his methodological introduction to *Argonauts of the Western Pacific*, he set forth his professional bona fides and procedures—his mastery of languages, time spent in the field, techniques of collecting and recording data—for readers to assess and judge. He warned readers that statements unaccompanied by concrete illustrations were not eyewitness evidence but only what informants told him—hearsay, and facts only provisionally. And when he later discovered lapses in his methodology, he was quick to set the public record straight; as when in *Argonauts* he too hastily categorized gifts of husband to wife and father to children in Western terms as "pure gifts." "I have fallen," he confessed, ". . . into the error . . . of tearing the act out of its context." In an appendix to his book on gardening and land tenure, he published a full account of all the "gaps, failures and mix-ups" in his Trobriand fieldwork.[46] Textual practices acquired as a novice philosopher of physics thus remained an integral part of Malinowski's ethnography. It mattered for his field science that he entered it through a library door.

INTO THE FIELD

Malinowski's fieldwork consisted of three excursions from his home base in Australia to the islands of eastern Papua and western Melanesia.[47] The first, from late September 1914 to late February 1915, was to the island chain along the south coast of eastern Papua, where he spent most of his time (about seventy-three days) on the tiny island of Mailu, with excursions west to Port Moresby, the administrative center of British Papua, and east to Samarai, the region's trade center and gateway to the Melanesian archipelagos to the north and east. His second excursion, and the first to the Trobriand Islands, followed an interlude of writing at Adelaide and kept him in the field from late June 1915 to late March 1916. His base of operation in the Trobriands was the chiefly village of Omarakana in Kiriwina, the main Trobriand island, with short periods in other villages. Following a second and longer interlude of almost two years, in Melbourne, organizing field data and writing, Malinowski returned to the Trobriands, where he resided—now in a less prestigious, nonchiefly village—from early December 1917 to mid-March 1918, with side trips to the Amphletts and other island groups. Michael Young calculated that Malinowski spent a total of twenty-five months among the Trobrianders (not the thirty that he claimed). And diaries and letters reveal a picture of residence punctuated by bouts of "surfeit of native" and temporary respites in the company of local Europeans, especially the pearl trader Billy Hancock, a close friend.[48]

The details of Malinowski's three trips are intricate, and it is not always clear exactly what he did or why. What is clear is that his first experience of actual fieldwork was a conventional ethnological survey. His principle mentor, Charles Seligman, remained committed to history and regional survey, and Malinowski was expected to become a collector of interviews and artifacts. Yet despite initial despondency and self-doubt as to his capacity for practical fieldwork—sharpened no doubt by memories of his earlier failure in the lab—by the end of his first voyage, he knew that he could do the work and do it well. As he wrote in his private diary, "I have been in N.G. [New Guinea], I have accomplished a good deal. I have prospects of far better work—fairly certain plans. And *so*—it's not as hopeless as I thought when I arrived here. . . . True, it's not all over yet; but in the light of old fears and uncertainties I have decidedly won a victory."[49] On the south coast of New Guinea, he regained his shaken confidence in his ability to produce original scientific facts—and not just in a library now, with texts, but in the field. He had a future as a working field ethnographer, that was clear.

Like the seven or so ethnologists in training who had preceded him to the South Seas, Malinowski was charged with filling in geographical gaps in surveys previously carried out by Seligman and Rivers.[50] Mailu, on the south coast of eastern New Guinea, was chosen as Malinowski's first stop because Seligman had failed to get there himself in 1903–4 and wanted evidence of cultural mixing of Papuan and Oceanic peoples. Malinowski's second trip, to the north coast of Papua, was planned to complement the survey of the south coast, concentrating on Mumbare, historically the center of radiation of important religious ceremonies. The Trobriand Islands were initially not on Seligman's itinerary at all, though Malinowski may have quietly planned a brief visit, having heard reports that the Trobrianders were an exceptionally energetic and aristocratic people, who made the best deep-water canoes and artistic objects, had the best dances, and—it was rumored—were adepts in witchcraft and sorcery. They were, in short, the sort of artistic, cosmopolitan people that appealed to the cosmopolitan Anglified Pole.

It was not until he was en route to the north coast of New Guinea and beyond recall that Malinowski wrote Seligman of his intention to proceed to Mumbare via a long detour north to the Trobriands. He was worried that Seligman would object to his going over ground that had already been worked in Seligman's earlier survey, so he devised various plausible reasons for his detour. He need not have worried and schemed. Seligman would never have objected: the only thing that mattered to him was getting good material, and Malinowski was soon getting very good material indeed in Omarakana. Besides, it was the custom in expeditionary work for planners at home to allow agents in the field to pursue unforeseen opportunities. Malinowski never did get back to Papua (though he kept saying he meant to). Nor did he make it to the farther destinations in Seligman's planned survey of islands to the east (Misima, Sudest, Rossel). He told Seligman that circumstances (e.g., vagaries of transport) forced him to give up their survey plan and remain in the Trobriands.[51] In truth, he wanted to stay and was aided by circumstances.

The subterfuge is characteristic of the man. Malinowski was temperamentally a neurotic worrier and schemer, imagining situations to be more complicated than they actually were, and making them more complicated for himself and others by scheming to shape situations to his purposes. He had the kind of personality that is labeled, euphemistically, "complicated": meaning brilliantly original and charismatic, but also troublesome and exasperating. Colonial authorities mistrusted him. They did not like a subject of the Austro-Hungarian Empire running around a British colony when the two empires were at war, and regarded him as an unpredictable and disagreeable nuisance. Unable to read his motives or intentions, they feared the worst: Was he an immoralist? An Austrian spy?[52]

Although Malinowski was no historical ethnologist, he did not go to the Trobriands thinking to pursue a different kind of ethnography. His first expedition and part of his second were in practice decidedly in the Riversian survey mode. In New Guinea he interviewed resident Europeans and native informants, starting at Port Moresby with Ahuia, Seligman's prize informant a decade earlier, taking his cues from Rivers's 1912 edition of *Notes and Queries*. His report is organized by categories right out of *Notes and Queries*—facts presorted into European pigeonholes.[53] It was what he was expected to produce, and a quick publication would impress his Australian hosts and sponsors and secure continuing support from London. He knew from Rivers's experience that he would get more complete and reliable information by interviewing natives in their villages and in their own language without other Europeans present.[54] But in a large-scale survey, there was a schedule to keep and no time for unhurried observing and exploring en route. It was improved empirical methodology that he sought, not a practice that went beyond intensive survey—not yet, anyway.

Malinowski later regretfully recalled suffering in his first trips from a belief in the "infallible methods" of Riversian positivist methodology, but the regret was a hindsight judgment. At the time he sought not just to follow Rivers's methods but to perfect them: "I still believed that by the 'genealogical method' you could obtain a fool-proof knowledge of kinship systems in a couple of days or hours. And it was my ambition to develop the principle of the 'genealogical method' into a wider and more ambitious scheme to be entitled the 'method of objective documentation.'"[55] The aim of this "objective documentation" was to treat every native activity (e.g., land tenure) with the same empirical rigor with which Rivers treated kinship. He was once more the clever and ambitious acolyte doing his empirically exacting mentors one better. In his first months in residence in Omarakana, in July and August 1915, Malinowski was practicing an intensive form of ethological survey, his faith in the empiricism of Rivers and Mach quite intact.

In the same spirit he drafted a thousand-page "ethnographic blockbuster" (Michael Young's term), laboring on it off and on between 1916 and 1918. This inventory of all the facts he had collected in the Trobriands was an embodiment of the ideal of grand ethnological survey and of Machian empirical rigor. In the end he never finished the book, but instead mined it for material for his series of thematic monographs on the principal social institutions of Trobriand society—a literary form that embodied the new practices of participant observing and functionalist sociological anthropology.[56] Intensive survey was thus a bridge from survey ethnology to resident observing and community ethnography. Probably there was no precise moment of metamorphosis. Malinowski began in one mode and ended in the other. That was how he described the experience.

RESIDING AND THE "IMPONDERABILIA" OF EVERYDAY LIFE

The Ethnographer arrived to take up residence in Omarakana in mid-July 1915: not on the beach, by boat, but less romantically by a ten-mile walk with carriers and sixty cases of gear from the west coast town of Losuia—Kiriwina's European administrative center and gateway—where a missionary steamer had dropped him off a few weeks earlier. Unable to engage in interviews or conversation until he learned the language, Malinowski concentrated at first on "objective documentation" in the Rivers style: taking a village census, collecting genealogies, and compiling a list of kinship terms, while gleaning what bits of information he could about village lives as they were actually lived and experienced.[57]

The "imponderabilia of daily life," Malinowski called these small and seemingly inconsequential observations. They were at first an incidental bonus of residence, but as he learned more, they became his most fruitful and valued evidence.[58] Malinowski's odd but apt neologism marks the shift from survey methods to resident observing. "Imponderabilia" did not arrive in familiar categories, like the facts produced in census or *Notes and Queries* interviews. Rather, their meanings had to be worked out if possible on the spot, from what the ethnographer knew of village customs and the contexts of particular occurrences. If these densely situated imponderabilia fell short of the epistemic standards of the generic and numeric facts of "objective documentation," they were in their way no less factual, and far more revealing of the operating principles of human societies. In them Malinowski could perceive "the charters of native institutions"—the communal laws and customs that allowed the expression of individual ambitions and desires while maintaining social order and giving daily activities shared meaning and legitimacy.[59] In these imponderabilia were the seemingly inconsequential clues that to the alert and knowing Sherlockian eye revealed patterns hidden in the actions and accidents of everyday life. They were ethnographers' inside stories.

As there is little for me to add in the way of facts to what is already known of Malinowski's participant observing, I will play the visiting historian and simply evoke and interpret what is known.[60] So, readers, imagine yourself suddenly set down on the beach in Omarakana, or arriving footsore from the west, looking for the Ethnographer, whom you have arranged to shadow and interview as he goes about his work. You may find him in his tent, with a knot of native men and children lounging just outside looking in and making loud and sometimes rude remarks in an unpleasantly jocular tone. Or you may catch up with him idling and gossiping with a group of Kiriwinans on the porch of a neighbor's hut; or in the gardens observing yam culture or harvest; or on a walk with a mob of chattering children in tow; or looking on discreetly with notebook in hand as neighbors dispute the rights and wrongs of some village drama. It's all quite inchoate. Your ethnographer is clearly in residence and neighborly, but where exactly is the science in this casual and indiscriminate looking and jotting notes? What are the operating principles of the tribe of resident ethnographers? Pitch your own tent nearby and, notebook in hand, follow the Ethnographer in his daily rounds.

Malinowski had pitched *his* tent—actually a borrowed tent: the one he brought from London had proved to be too small—in the high-status ring of huts near the hut of the chief, To'uluwa, and not far from that of the garden magician, Bagido'u, who became a favored informant and particular friend. Though not yet fluent in the Kiriwinan language—he had planned only for a short stay—he learned languages easily and within a few months was interviewing his neighbors in their own vernacular without an interpreter and with only an occasional lapse into pidgin. More and more he was taking notes of conversations in Kiriwinan, at first just important phrases and eventually everything. In time he could follow and record Trobrianders' conversations among themselves; that milestone was passed early in his second period of residence. The difficulty was not vocabulary and syntax, which in Kiriwinan are very simple, but catching the subtexts and unstated meanings of what was said. That required knowledge of the whole context of village personalities and relationships.[61] To use the language he had to know the society as well, and that took time.

The facts produced by resident observing were unlike those produced by formal interview and were generated and recorded in a different way. Imponderabilia were, for one thing, repetitious and redundant. Malinowski made it a cardinal rule of his field practice to record everything he heard and saw, however trivial or repetitious it might seem. He filled notebooks with whatever he happened to hear and see: new facts and old, and the views of all villagers whatever their age, gender, or social standing.[62] Such unfiltered data were harder to organize than the presorted data from standard questionnaires; but they had unique advantages. They revealed, for one thing, the great variation in individuals' views and feelings, and how seemingly fixed social conventions operated not as absolute laws but as flexible and evolving guides to conduct in particular situations. The total data of resident ethnography revealed a community's social life not as it was supposed to be lived but as it was lived in fact.[63] An analogous transformation occurred in systematic biology when taxonomists began to collect repetitiously and redundantly—not a few "typical" specimens of each species but hundreds, from many places—and to keep ecological field records. In so doing they changed a static, typological conception of species into a populational and evolutionary one.[64] So likewise did the gather-all method of resident ethnography change a typological view of human societies into a diversely sociological one.

A second way in which the facts of resident observing differed from those of interview and survey was their seeming disorder. Malinowski set down his observations not in European categories of economics, religion, family, and so on, but in chronological order, as it all happened. His notebooks were an ethnographic diary or cinema verité. Michael Young remarks on "the surreal concatenation of subjects in his notebooks," and the "kaleidoscopic effect" of "imponderabilia galore." Malinowski made no effort to select or highlight significant events. "[T]here is no hierarchy of topics," he wrote. "Nothing is too trivial to record. Every sensation, every observation, is grist to the diarist-ethnographer's mill."[65] It was not disorder, however, but the inchoate order of social life, as Malinowski later explained to his assistant, Camilla Wedgwood, in New Guinea. She was dismayed by the disorder of her field data. But her records, he assured her, were excellent:

> This I want very much to impress on you: field-work in its best form looks just like the thing which you are sending us over. It must at first be chaotic, and put in the form of little odds and ends. One gets whole weeks of complete disillusionment and despondence. And then suddenly after months of toil and labour one or two institutions suddenly fall into focus and one or two strokes allow us to build up the full picture. And this is the joy of field-work. But you must not expect it to happen immediately.[66]

These humdrum changes in practices of data management had far-reaching consequences. Because the recorded data were ordered in the patterns of daily life, they gave access to categories of activity that were Trobriand, not European, and to associations between activities that were erased in the artificial order of stock verandah interviews. Records of lived experience from the inside revealed that Trobriand social activities were not separate and specialized as in European society, but composite. Kula trade and gardening, for example, did not just provide trade goods and food, but also sustained principles of reciprocity and matriliny vital to communal solidarity. Fishing, to which Trobrianders were devoted and which was elaborately enmeshed in magic performances, was equally foraging and religious practice.[67] Data recorded in the form of events unfolding in village life preserved the social situations of speech and the connections between speaking and acting that gave talk meaning and force and were lost in the detached situations of formal interview.

Sitting at the ethnographer's table and answering preset questions, informants naturally adopted the appropriate demeanor of village authorities called upon to display their knowledge of the conventional view of things, what "everyone knew." The performative situation demanded a correct and authoritative performance. The contexts of daily life, in contrast, demanded words and actions appropriate to the particular situations and dramas that were unfolding. As Malinowski put it,

> When the native is asked what he would do in such and such a case, he answers what he *should* do; he lays down the pattern of best possible conduct. When he acts as an informant to the field-anthropologist, it costs him nothing to retail the Ideal of the law. His sentiments, his propensities, his bias, his self-indulgences, as well as tolerance of others' lapses, he reserves for his behaviour in real life. . . . The other side, the natural, impulsive code of conduct, the evasions, the compromises and non-legal usages are revealed only to the field-worker, who observes native life directly, registers facts, lives at such close quarters with his "material" as to understand not only their language and their statements, but also the hidden motives of behaviour, and the hardly ever formulated spontaneous line of conduct. "Hearsay Anthropology" is constantly in danger of ignoring the seamy side of savage law.[68]

Setting down observed facts in full and in the order they were lived situated them in the contexts of actual life, thereby enabling ethnographers to read their social meanings as participants did. That context was preserved in the seeming jumble of verbatim field notes.

Malinowski the resident observer became a vehement critic of survey-style interview—"hearsay anthropology," he called it.[69] It was the term he had used in his book on the Aboriginal family for published facts that had no supporting evidence of who, when, where, and how. He likewise disparaged the encyclopedic ethnological monograph as the "litany method," which inventoried atomistic facts in the same way that ethnographic artifacts were once lined up in museum cases with no contexts of use and meaning The desituating empirical practices of ethnological survey were no more respect-worthy in his view than the facetious ignorance of the "old recorders"—travelers, traders, missionaries—who, not understanding the context and meanings of what they witnessed, represented it as comical or "queer."[70]

Recording the imponderabilia of village life entailed full-time, sustained engagement of the ethnographer with his subjects—the engagement of residing. Simply being there physically was not enough. It was the social relationships of residence that made it a scientific practice, and relations had to be established and maintained. Malinowski became a resident as traders and missionaries did, but one who lived among his subjects, not in European enclaves, and devoted himself exclusively to understanding and recording his neighbors' way of life.[71] He developed a daily routine of drifting and hanging out with no particular plan or agenda. He talked not just with designated informants or village leaders but with everyone, and not just men but women and children. An advantage of Omarakana was that village women were not shy of whites, as they were in many villages. As a resident and neighbor he began to experience village life as residents did themselves:

> Soon . . . I began to take part, in a way, in the village life, to look forward to the important or festive events, to take personal interest in the gossip and the developments of the small village occurrences; to wake up every morning to a day, presenting itself to me more or less as it does to the native. I would get out from under my mosquito net, to find around me the village life beginning to stir, or the people well advanced in their working day. . . . Quarrels, jokes, family scenes, events usually trivial, sometimes dramatic but always significant, formed the atmosphere of my daily life as well as of theirs.[72]

Following each morning's ethnographic labors in the village, he would take a long afternoon walk in the surrounding country, often with a gaggle of kids in tow. These proved to be some of his best sources of inside information: "[W]ithout the constraint of being obliged to sit and be attentive, they would talk and explain things with a surprising lucidity and knowledge of tribal matters. In fact, I was often able to unravel sociological difficulties with the help of children, which old men could not explain to me. The mental volubility, lack of the slightest suspicion and sophistication, and possibly a certain amount of training received in the Mission School, made of them incomparable informants."[73]

Malinowski learned most when his informants were not on their guard or nudged by the situation of official questioning to say the correct, official things. Caught up in some exciting event—a family quarrel or breach of custom, the unexpected arrival of visitors from the sea, preparations for a feast or ritual or kula voyage—villagers would be "too excited to be reticent, and too interested to be mentally lazy in supplying details." Conversations with villagers engaged in activities that mattered to them—gardening, fishing, making tools or clothes, relaxing, playing games—yielded facts that were not the official reality of Trobriand society but the unofficial actuality—how things really worked.[74] As a resident stranger Malinowski enjoyed the dual advantages of intimacy and distance. He was not strictly bound by Kiriwinan rules and customs; so when he unwittingly committed breaches of etiquette, as he often did, his neighbors generally did not take offense but simply laughed or mocked and set him straight. Even gaffes could be turned to science.

Sometimes Malinowski would put aside the tools of his own resident occupation—notebook, pencil, camera—and enter actively into children's games and the minor routines of adult life, partly for his own education and partly just for human companionship. These "plunges into the life of the native" were easier for him as a Pole than for western Europeans, he thought, because the Slavonic nature is more "plastic and more naturally savage." What exactly he meant by that is not clear; but his words bring to mind that citizens of the polyethnic Austro-Hungarian Empire had to be able to operate in varied cultural contexts, whereas citizens of ethnically homogeneous nation-states had no such need. In any case, his "plunges" into participation were the moments in which he felt closest to understanding Trobriand society as Trobrianders themselves did.[75]

His immersion in village life was not total, however. As Raymond Firth put it, "[H]is participation was almost always secondary to his observation." In his relations with his subjects, "his cerebral functioning had almost complete ascendency over his 'gut' reactions of involvement."[76] Malinowski took no part in any of the major activities of which he wrote: baloma and other ceremonies, kula voyaging, sex and family life, clan politics, gardening. He did on occasion allow himself to be drawn into dancing but generally observed ceremonies inconspicuously from the sidelines. It was too easy as a participant to unwittingly offend, as he once did by paying for new dances to be performed in a baloma celebration, which greatly displeased the baloma spirits, who in retaliation did mischief in the village.[77] He wanted very badly to experience a full cycle of kula voyaging but was banned from taking any further part after a short excursion, when ill winds earned him the reputation of a bringer of bad luck and witches—a Jonah. After that kula crews departed surreptitiously, to make sure the Ethnographer stayed behind.[78] He did not kayta—have sex—though he was sorely tempted to take part in what all Trobrianders freely and openly enjoyed from a young age.[79] He was allowed to assist in small ways in Bagido'u's garden magic, but he took no part in the clearing, weeding, and digging; and he did not pile up yams in front of his tent to display skill and social standing, as was the custom.[80] He took pains to avoid being identified with any village political faction, lest he make enemies of all the others. In public conflicts he remained strictly a spectator, notebook and pencil in hand, asking questions but not taking sides. The anthropologist Ruth Fink, in her taxonomy of participant practices, categorized Malinowski's as "incomplete participation."[81] Participant observing was a double life: engaged and disengaged, always in but never fully of the community, local and cosmopolitan, a little strange but no stranger.

In fact, Malinowski was playing the role that villagers expected of him as resident ethnographer. Omarakanans remembered him as "Tolilibogwa," the man of myths or legends, or "man of songs"—a role familiar to them. "The Historian" would have done as well, Michael Young suggests.[82] I'd say the Trobrianders got the residential identity of their odd new neighbor about right, certainly more right than many white colonials did, who disparaged anthropologists generally as "anthrofoologists" and could see no sense or use in what they did.[83] The efficient resident magistrate of Kiriwina, Raynor L. Bellamy, developed a strong dislike of Malinowski, among other reasons for his encouraging of traditional customs that Bellamy the Westernizer was doing his utmost to stamp out.[84] The missionary and amateur ethnologist William Saville, with whom Malinowski had stayed (and quarreled) in Mailu, thought him incapable of entering into the minds and situations of natives or of treating them as people much like himself.[85] Malinowski's neighbors in Omarakana were more understanding, assigning him an identity that fit his actions and made sense to themselves, and made a living and working relationship possible.

Malinowski never claimed to have an emotional "gut" empathy for the people he studied, though that was imputed to him by a later and more romantic generation of ethnographers, who were then outraged when his private diary was published and dispelled that wishful view. For example, in a moment of "surfeit of native" and of intense yearning for his beloved Elsie Masson back in Australia, Malinowski confided that "[a]s for ethnology: I see the life of the natives as utterly devoid of interest or importance, something as remote from me as the life of a dog." He was not saying that Trobrianders were as worthless as animals, as some commentators believed; only that their way of life and his own were in the end worlds apart. In fact, Malinowski respected Trobriand culture and thought it in no way inferior to that of Europeans, and disdained European colonials who presumed cultural superiority.[86] His role as resident recorder and storyteller was not to empathize or advocate but to observe, record, and understand Trobrianders' way of life in their own, not European, terms. For that, empathic feelings and cultural identification were not necessary; indeed, they would get in the way.

It was the accepting yet objective practices of residing and observing that enabled Malinowski to grasp the operating principles of Trobriand society. The "Ethnographer's Eye," he wrote, is like "a super-cinema operating from a helicostatic aeroplane" hovering above the village, observing how it was all structured and interrelated.[87] It is a strikingly kinesthetic image of resident observing. "Tolilibogwa," the ubiquitous nosy neighbor and collector of stories, combined the ground-level view of a resident with the aerial view of cosmopolitan science. He aimed to understand his subjects both as an ethnographer and as they understood themselves: "To discover what are his [the native's] main passions, the motives for his conduct, his aims . . . [h]is essential, deepest way of thinking."[88] Villagers had no such helicostatic view, only the view from the ground.

Malinowski once likened his habituation to Trobriand life to his experience as a Pole assimilating into English society. In a letter to Sir James Frazer, he observed that for a foreigner in England to comprehend English institutions he must know how to live in them: to take part in sports and amusements, school and university customs, politics.[89] This might seem to suggest that his experience as an immigrant Pole in London was a dress rehearsal for participant observing. However, the two experiences were sociologically quite different. In England Malinowski meant to become a fully participating citizen—to go native. In Kiriwina going native would have put an end to the science. He was, besides, never really an alien in England. An Anglophile from an early age, he was drawn to England in part, as he said, by his admiration of English upper-class manners; and as a student at the London School of Economics, he gravitated to the most cosmopolitan of the English ethnographers, Charles Seligman (a German Jew) and Edvard Westermarck (a Swedish Finn). No Polish nationalist (he never wanted to repatriate), Malinowski was by upbringing and education a citizen of an international European republic of letters and science. It was in Omarakana that he was first challenged to live and think his way into a culture that was truly foreign to him.

RESULTS AND DISCOVERIES

Malinowski never set down exactly when and how the power of resident observing was revealed to him. However, his Trobriand monographs are full of first-person stories of episodes in which that power was vividly manifest in what he learned and how. These stories were not meant to be history, obviously. Yet they do reliably reveal how participant observing opened unexpected insights into the structure and operating principles of village society. In some cases abundant new facts revealed the meaning of puzzling actions and events. In other cases an unsettling word or small clue blew up generalities that ethnographers had believed to be beyond challenge. In his first weeks in Kiriwina, Malinowski experienced mainly the thrilling richness of material that in New Guinea he had glimpsed as a future possibility. Before long, he was writing of finding "lots of stuff which absolutely smashes my views about kinship and proves the imbecility of some of my seriously and elaborately constructed views."[90] After years of exposing errors of fact and reasoning in the work of others, here he was, having to own up to his own methodological failings. That shaking of bedrock ethnological verities was the context in which resident observing could be perceived not just as an improvement in survey practice but as something distinctly new and better.

One such revelation concerned the vital matter of procreation. It was a fact universally accepted by ethnologists at the time that native Australasians (and many indigenous peoples) were ignorant of the physiology of reproduction: women made babies, men had no role in it, intercourse did not cause pregnancy.[91] The issue was vital because it was the bedrock of ethnologists' obsessive concern with kinship. Ignorance of paternity sustained customs of matriliny and open sexuality, just as knowledge of paternity sustained European customs of patriliny, patriarchy, and patristic religion. Informant testimony overwhelmingly backed the received view. However, informants did occasionally hint of a connection between coitus and conception: in casual remarks that virgins did not conceive but cohabiting women did, and that a month of regular intercourse was required for conception. Yet when directly pressed, they steadfastly denied that sex caused pregnancy. Such contradictions were baffling, but because they were rare did not dispel the ethnological consensus. Yet for someone devoted to empirical rigor, like Malinowski, they were a nagging loose end; so when he took up residence in Omarakana, he asked nagging questions.

In the Trobriands the issue of conception was bound up with a dense knot of beliefs about afterlife and rebirth. Trobrianders believed that the spirits of the newly dead, or baloma, journeyed to Tuma, an island just to the north of Kiriwina, where they lived their afterlives through cycles of death and rebirth. Once a year the baloma voyaged to Kiriwina to be among the living and to make sure that the dances and ceremonies in their honor were correctly carried out, and to cause trouble—falling coconuts, bad weather, contrary winds—if they were not. The baloma also caused pregnancies, by ferrying the spirits of unborn children from Tuma and inserting them into women, either through the forehead or the vagina (accounts differed). Unmarried girls avoided swimming at high tide, lest among the flotsam there lurked a baloma waiting to insert a spirit child. So when Malinowski asked directly what caused a woman to conceive, he was invariably told that a baloma brought the child.

At first Malinowski felt secure in his adherence to proper ethnographic methods. He had "obtained this information quite smoothly" by personal interview of trusted informants, so there could be "no more difficulties to be cleared up." The more questions he asked, however, the less certain he became. He was told that a loose woman was more likely to have a child, and that a virgin, if one could be found (girls began an active sex life as young as six or eight), could not get pregnant. Thus "prompted by the instinct of pure pedantry," he revealed "a flaw in the very foundations of my construction," which "seemed threatened with complete collapse." It was not just the standard story that was shaken, but the empirical methods from which it was derived. "[T]he contradictions and obscurities . . . appeared to me quite hopeless," he recalled. "I was in one of those desperate blind alleys, so often encountered in ethnographical fieldwork, when one comes to suspect that the natives are untrustworthy, that they tell tales on purpose; or that one has to do with two sets of information, one of them distorted by white man's influence."[92] As the method of informant testimony became more searching and exact, it began unexpectedly to fail.

The "final shock" to his trust in the conventional view was hearing the story of how a mythic hero, Tudava, was conceived. His mother, Bulutukua, lived alone in a grotto and one day fell asleep, and water from a stalactite dripped on her vulva, opening her passage so that she conceived the hero, and later several more children as well, without the dripping water. Little hoping to discover any logical sense in all this, Malinowski nonetheless persisted and was finally rewarded with the solution, which was in fact quite simple: intercourse was necessary to conception but only mechanically, to open the vagina so that a baloma could insert a spirit child. Further questioning proved beyond doubt that the idea of mechanical opening was no aberrant opinion but what everyone believed, yet to Europeans stoutly denied.

Malinowski learned as well, though not until his second stay in the Trobriands, why his informants had denied so vehemently that intercourse caused conception: it was because the missionaries had been pressing them to adopt that belief, and with it the whole unwelcome package of European sexual morality and patristic religion. Malinowski's neighbors were angry that their historian, usually so unmissionary, would persist in talking missionary nonsense. His eyes opened, Malinowski then began to ask them about "missionary" views, and that way of putting it freed them to state openly their real beliefs. One old man demonstrated the self-evident truth by opening and closed his fist, showing how a baloma could easily insert a child into the one but never into the other.[93] Beliefs that had seemed irrational to ethnologists were thus revealed to be perfectly logical and, in the context of matrilineal institutions of clan and family, more reasonable than the missionary account.

Malinowski's novel solution to a long-standing puzzle was a sign of his growing ability to think Trobriand. Resident observing was giving him imaginative access to Trobriand categories and social logic, and imaginative freedom from his own European ones. In grasping the logic of Trobrianders' view of conception, he made a categorical distinction between mechanics and physiology that was as unthinkable for Europeans as the missionary nonsense of insemination was unthinkable for natives. In effect, Malinowski resituated the issue from the mechanistic world of European science into Kiriwinans' world of social and symbolic facts—a resituating made possible by sustained resident observing.

The baloma spirits provided the occasion for another demonstration of the power of resident observing, in the ceremonies and accompanying feasting, dancing, sexual license, and general rejoicing that followed the annual harvest. The baloma were an important presence in this ceremonial cycle, checking up that gardeners had worked well, and that ceremonies were properly done. Sailing in from Tuma, they landed on the beach on an incoming wind or tide and lived either with their maternal relatives or on the beach by their canoes, as any visitors would do. They were honored with displays of valuables and yams and provided with food for sustenance. And their potential displeasure with gifts and ceremonial performances was taken very seriously indeed, as they were capable, if displeased, of instigating droughts and crop failures.[94] So in ceremonies villagers showed great respect to their unseen visitors. In their everyday demeanor and manner of speaking about the spirits, however, villagers gave a quite different picture. Malinowski discovered to his surprise that villagers' attitude to the baloma could be distinctly unceremonious, even disrespectful. There was no fear or avoidance taboos, and much ridicule and joking at the expense of the spirit visitors, who were expected to leave promptly as the ceremonial period drew to a close. He had expected villagers to display the same pious and superstitious fears that Europeans felt about their ghosts and spirits. The disparity between official and everyday attitudes was to a European eye bafflingly incongruous.[95]

This incongruity was most dramatically revealed when Malinowski decided to witness the final and essential act of the ceremonial cycle, when boys with beating drums ushered the baloma to their boats and saw them off: "[I]n no instance, perhaps, of my field work, have I had such a striking demonstration of the necessity of witnessing things oneself, as I had when I made the sacrifice of getting up at three in the morning to see this ceremony. I was prepared to witness one of the most important and serious moments in the whole customary cycle of annual events, and I definitely anticipated the psychological attitude of the natives towards the spirits, their awe, piety, etc."[96] What he saw, in fact, was five unruly and unsupervised kids doing what ceremony required, but in the brazen manner of street urchins: making facetious remarks, begging for tobacco, or performing "some nuisance sanctioned by custom." No detail of the prescribed ceremony was omitted, yet in its performance there were "no traces of sanctity or even seriousness." The baloma spirits were "ceremonially but unceremoniously driven away."[97]

What Malinowski discovered was the simple yet unexpected truth that Kiriwinans regarded and treated the baloma much as they would any visitors from another island: respecting the customs of hospitality and giving a good show of their own wealth and virtues, but with no neighborly or personal engagement. They were glad to see the visitors arrive, and glad to see them go. Unlike European ghosts, baloma were essential, if sometimes demanding and troublesome, participants in the human world. Such an insight could only be gained by participant observing. Survey ethnologists, who investigated only the showpiece ceremonies, would miss what made Trobriand conceptions of the spirit world so distinctive and illuminating of human social behavior—so different from European conceptions yet so understandable. It was the inside clues—the imponderabilia—that revealed the perfectly reasonable logic in what to European eyes made no sense.

The power of resident observing was similarly revealed in Malinowski's efforts to untangle Trobrianders' baffling rules of gardening and land tenure. He learned little from formal interviews. In New Guinea he pumped informants about legal ownership and inheritance, but they seemed baffled by the questions and gave bored, perfunctory answers. The problem, Malinowski soon realized, was that Melanesian and European concepts of property and ownership were so different that questions put to informants in European categories were unanswerable nonsense.[98] Yet in his conjectures he clung to the principles of European property rights: perhaps no one owned garden plots, or the chief owned it, or the community as a collective—the stock categories of European ethnoeconomics.

The Trobriands proved a more fruitful ground for investigating land tenure than Mailu, which was a trading society, whereas Kiriwinans were famously skilled and passionate gardeners, especially in the fertile area around Omarakana. As a resident Malinowski could observe activities that occupied villagers much of the time and about which they talked incessantly and passionately. He was also able to observe the whole seasonal cycle of performances of yam gardening: clearing ground, planting and weeding, harvest, display, and distribution. He was thus enabled to situate talk of gardening and tenure rights in material contexts of use, and to compare official accounts with real behavior. The rules of garden tenure proved to be more intricate—and more rational (in Trobriand rationality)—than ethnologists had thought.

Malinowski's road to understanding was, he recalled, "an odyssey of blunders in field-work."[99] And in that blundering we catch glimpses of how resident observing developed in situ out of intensive survey. Malinowski assumed at first that rigorous gathering of generic facts would reveal the principles of tenure rights, as it had revealed to Rivers the tangled principles of kinship. Applying his "method of objective documentation," Malinowski mapped gardens and fields, recorded who owned and worked what fields, compiled lists of native terms for the various forms of legal title and ways that land was inherited. From these data he identified all the categories of people who had a right to work a garden plot or to share in the harvest. These were the district chief, the village headman, the garden magician, the head of the subclan, the subclan as a whole, the village community as a whole, individual members of the community, the actual gardener, and the gardener's sister and female relatives. These last were crucial, because by the laws of matrilineal inheritance men were obligated to give most of what they grew to their sisters' households in distant villages.[100] But this list of categories proved to be as far as Malinowski could get with "objective documentation."

Malinowski's roster of stakeholders accorded with what informants told him, yet it left him quite in the dark as to how the institution of land tenure actually worked. The bare facts shed no light on the moral basis of claims, whether they were discordant or had some overall logic, and how conflicting claims were adjudicated. Nor was there any prospect that further inquiries, however rigorous and cross-checked, would reveal the social logic of usufruct. Asked why they did things as they did, villagers invariably replied, because that's the way we've always done things. The meanings of "why?" were culturally specific: for Europeans the question called for an account of cause-and-effect relations between isolated variables; for Trobrianders it called for an account of custom and history. "Objective documentation" produced abundant facts, but without living context ethnographers could not know what these facts meant. To drive the methodological point home, Malinowski invited readers of *Coral Gardens* to close the book, list the facts they had just read, and ask themselves what if anything they understood; not much, he clearly expected. Readers could thus recapitulate as virtual participants Malinowski's discovery of the failings of a practice that gathered bare facts lifted from the webs of social activity that gave them meaning in Trobriand life, but no meaning at all in the abstract categories of European science.[101]

It was Malinowski's practice of recording everything he observed as it happened that resituated the facts in those webs of activity. It was, as he put it, the "sheer chaotic welter of observed details" that revealed how gardening was integrated with other activities of Trobriand life.[102] Because he wrote down facts not in preset social science categories but as lived experience, activities that were associated in life but separated in science remained associated in the factual record. Activities that in European culture were distinct and separate retained in Malinowski's jumbled notebooks the connections they had in Trobriand life. Connections that might not be apparent in the welter of daily living stood out clearly when the written record was examined and reexamined at leisure.

Resituating the facts and lists of gardening lifted the veil from the baffling intricacies of land tenure. Omarakanans pursued gardening so passionately not just because they liked to be securely provisioned, but also because the activities of provisioning were integral to their social order and identity. Gardening embodied the foundational principles of reciprocity and displayed the communal virtue of responsibility to kin and community. Being a visibly successful gardener was how villagers earned respect and social standing. The virtues of good gardening were those of good citizenship—as were the virtues of making fine canoes and shell ornaments, and skillful blue-water navigating.

Customs of displaying and distributing the yam harvest were especially revealing of the connected meanings of provisioning. Malinowski watched and wondered at the custom—nonsensically inefficient to European eyes—whereby each gardener delivered a large part of his harvest of yams to sisters in distant villages only to receive in return equal amounts from those sisters' husbands. He observed how lovingly and obsessively gardeners piled their most perfect yams first at their gardens and then in front of their huts, leaving them to be admired and discussed even as they became unfit to eat. He observed the use of food in village ceremonies and gift exchanges, and the deference paid to successful gardeners and the passion with which villagers labored to earn that status. In producing, exhibiting, and giving yams Omarakanans pursued individual ambitions while maintaining communal solidarity. That was the social logic that lay hidden in Malinowski's "objective" list of stakeholders.

Malinowski identified four distinct and overlapping principles that guided claims to land and food. One was citizen or autochthonous rights:

these belonged to all members of the village subclan and derived from the myth of origin, in which the first brother and sister pair emerged from the ground at a particular place and laid claim to it for their descendants. Resident rights were a second kind of claim. These derived from the matrilineal requirements that wives at marriage move to their husbands' villages and that their sons at puberty move to their mothers' natal villages (contact between brothers and sisters was taboo). Obligations of family provisioning could thus not be fulfilled locally and efficiently—as Europeans would, who made a moral virtue of efficient production—but only by circulating staple food between paternal and maternal villages. Overriding both citizen and resident rights was a third principle, of social hierarchy and deference to the village chief, who received gifts of food and was expected in turn to provide abundantly for communal feasts and to villagers in need. And, finally, there was the claim of the garden magician, without whose precise knowledge of the necessary spells all gardening would fail. This intricate system of tenure rights and usufruct was a moral, economic, and social system of matrilineal social relations, in which yams were the material currency. Gardens asserted a community's right to live where they did, in perpetuity. Skill in gardening demonstrated communal values of self-sufficiency and reciprocity.[103] All this became clear when gardening was observed in action and situated in the web of activities from which it had initially been removed by the method of "objective documentation."

The working of native law is my final example of how the power of resident observing was revealed. For European visitors the law of "savage" societies had always been a puzzle. Resident "old recorders" who lived close to natives and watched the noise and tumult of their communal life saw no law and order at all but only the impulsive expression of individual passions and desires. In contrast, the first ethnologists saw *only* law and order—because that was all that their official paid informants told them: that laws and customs were absolute, with no leeway for individual or social passions. Resident observing dispelled both these views. In fact, native societies were ruled by laws and customs that in principle could not be questioned but in practice allowed for accommodations to the knotty particularities of human life.[104] Open breaches of law proved especially revealing.

A dramatic manifestation of the conflict between law and human passion occurred not long after Malinowski arrived in Kiriwina. The conflicting principles in this case were what Malinowski called Mother Right and Father Love. In a matrilineal society a man's possessions and public privileges descended not to his wife's sons but to his maternal nephews; and this law of Mother Right was backed by ancient observance (the way we've always done it) and by the "fact" that children were not the husband's flesh and blood but the mother's alone. On the other side was Father Love: the emotional reality of patrilocal family life, in which husbands cared for the wife's children to the time of puberty, when sons left to live with their uncles in the villages of their mother's birth. Inevitably husbands became more emotionally attached to the sons of their immediate family than to their nephews, who joined the family circle only at puberty. Mother Right had the force of law behind it, and if contested it prevailed. Father Love was legally weaker but had behind it the full force of human passion and interest (in the case of chiefs and other big men, especially, wealth and high office were in play). The result was a web of informal usage—a kind of village common law—that made evasions of the law, within limits, an accepted yet contestable fact of social life. For the most part this web of secondary law and custom kept Mother Right and Father Love more or less at peace. It was the occasional failures that most clearly revealed to resident observers the faulted moral bedrock of Trobriand society.

That was the situation in Omarakana when Malinowski arrived, to find an open and festering feud between the chief's favorite son, Namwana Guya'u, whom he had kept in the village in breach of the law, and his nephew, Mitakata, whose legal rights were thereby infringed. The feud came to a head when Mitakata was jailed as the result of a litigation by the son, an injustice of which villagers disapproved. As night fell on the day that the news arrived, the central place emptied as families retreated to their huts, full of foreboding. The quiet was broken by the loud piercing voice of Mitakata's eldest brother, denouncing Namwana Guya'u as a troublemaking intruder and abuser of village hospitality, and a transgressor of the law of inheritance. In proper ritual language he demanded the son obey the law and leave: "We do not want you to stay here. This is our village. You are a stranger here. Go away! We chase you away! We chase you out of Omarakana." Thus called out, the son had no choice but to leave and that very night slipped away to live in his maternal village. This did not end the feud, however. When Malinowski returned to Omarakana two years later, he found the chief's wife dead (of grief, it was said), the chief withdrawn, the nephew's marriage broken (his wife was of the other clan), and the rival families more at odds than ever.[105]

The confrontation that Malinowski observed in the dark and deserted village—"one of the most dramatic events which I have ever witnessed in the Trobriands," he recalled—brought him "face to face with the discrepancy between the ideal of law and its realization, between the orthodox version and the practice of actual life."[106] Like the sight of the baloma being shooed rudely out of town, the showdown between nephew and son revealed to Malinowski what public ceremonies and formal interviews never could: "human cultural reality is not a consistent logical scheme, but rather a seething mixture of conflicting principles."[107]

Another such event revealed analogous informal accommodations to the law of clan exogamy. A few months after his arrival, Malinowski got word from a nearby village of an outbreak of wailing for the death of a boy he knew, who had fallen to his death from the top of a palm tree. Mortuary ceremonies were in progress when Malinowski arrived, and so intent was he on writing down every detail of the ceremony—his first—that he failed to note some odd features of the situation: a second youth, who had mysteriously suffered unexplained injuries; and palpable hostility between the village in which the boy had died and the one in which he would be buried. It was only much later, when Malinowski knew the full context, that the meaning of these clues became clear.

The boy who died had not fallen from the palm: he had jumped. He had transgressed the law of clan exogamy by romancing a maternal cousin. In a matrilineal clan line, first cousins were by law brothers and sisters, so that any sexual intimacy between them was incest, and tabooed. Nonetheless, maternal cousins were attracted to each other, and accommodations were made to unruly human passions. As Malinowski gradually learned, there were sanctioned ways of circumventing the law of exogamy: magic that, if performed appropriately and exactly, would undo the ill effects of intraclan incest by canceling the force of the law of exogamy in particular cases. (There was, as well, magic for undoing the magic of evasion.) So long as a couple was discreet and obeyed the customs of evasion, their crime was tolerated, even though it was known to all. If they provoked a scandal or were publicly denounced, however, the law was applied without recourse. To remain in the community after public exposure would have been intolerably shameful, and the only options were exile or suicide. Publicly called out by the girl's discarded lover, the errant boy had no choice but to climb the tree and jump. However, there were also limitations imposed by custom on those who upset the status quo of tolerance. So when the boy, high in the tree, cried out to his clansmen to avenge his death, his relatives could not refuse; it was in the ensuing fight that the accuser was seriously wounded, and the maternal and paternal villages irreparably sundered.[108]

These intricate systems of law and accommodation could be discovered only in the imponderabilia of resident observing, never in the accounts of formal interviews or lists of objective documentation. Survey ethnographers, by removing informants from their social situations into ethnographers' scientific space, isolated themselves and their subjects from the contexts essential to understanding. Taking up residence in Kiriwina as the village observer and recorder, Malinowski in effect restored subjects of scientific study to their natural contexts. He made the village a venue of situating science, inserting a new social activity—ethnography—into the web of village activities.

CONCLUDING THOUGHTS

I referred in the first pages of this chapter to Malinowski's dual scientific persona: the great empirical field-worker and the uninspired general theorist. In his Trobriand monographs there is no such dichotomy. Malinowski does interpret, abundantly, but his interpretations do not derive from formal theory but are embedded in narratives of the concrete particulars—the "imponderabilia"—of daily life. This union of empiricism and reasoning has not gone unnoticed. Malinowski's student Edmund Leach got it about right: "That Malinowski was an imaginative genius of a high order there can be no doubt," he wrote, "but he had a bias against abstract theory which kept his imagination firmly earthbound. The result was a unique and paradoxical phenomenon—a fanatical theoretical empiricist."[109] That seemingly paradoxical term—*theoretical empiricist*—neatly captures the distinctive character of Malinowski's resident science. The sociologist Talcott Parsons made a similar point when he likened Malinowski's field practice to the empirical case method of clinical diagnosis and science.[110]

This insight has in fact been expressed in many ways. Carlo Ginzburg's Sherlockian method of clues and conjectures is another view of theoretical empiricism, and Samuel Gorovitz and Alasdair MacIntyre's idea of a science of particulars, yet another. So, I think, is the anthropologist Clifford Geertz's famous concept of "thick description," which despite its name is not description but a way of making clear the customary order of a society through a narrative of observed particulars rather than by reference to general social theory.[111] It is, in other words, a kind of theoretical empiricism. And digging deeper in the past, it is what the poet and naturalist Johann Wolfgang von Goethe was getting at, I think, when he wrote approvingly of "a delicate empiricism which so intimately involves itself with the object that it becomes true theory."[112] The anthropologist Adam Kuper voiced the same thought in almost the same words when he wrote that "[b]ehind Malinowski's concern with field methods . . . there was a grasp of the complexity of social reality which amounted, almost, to a theory."[113] Malinowski's "theoretical empiricism" was a practice of situated knowing that was both new and very old. And, arguably, it is characteristic of any science that is carried on inside its objects of study.

Malinowski himself always insisted that his empirical science was essentially theoretical, just not formally or dogmatically so. Theorizing was for him an active symbiosis of gathering facts and "theoretical moulding" that turned a jumble of atomistic data into the "invisible facts" of social institutions and their operating principles.[114] There were no such things as pure facts to be lifted out of worldly situations and examined: the Riversian method of "objective documentation" was in the end more likely to obscure than to illuminate. Social facts were made in a running dialogue of observing and reasoning in contexts of social choice and action.[115]

Although Malinowski developed his theoretical empiricism in the field, the basic idea had deep roots in his youthful determination to live in a way that would integrate positive science with artistic and literary imagining. We see that impulse in his early insistence that "metaphysics" was not inimical to natural science, as Mach asserted, but essential to it. He meant not the old metaphysics of systematic philosophers, but a metaphysics of material facts. "Substantive metaphysics" he termed it—another aptly paradoxically term.[116] It was a metaphysics that fed neither on pure cerebration nor on poetic fantasy but on an intimate and inquiring knowledge of the particulars of human society or nature. That ideal was what drew the young Malinowski so strongly to the writings of James Frazer. Frazer, he wrote in an early critical appreciation, "was endowed with two great qualities: the artist's power to create a visionary world of his own; and the true scientist's intuitive discrimination between what is relevant and what adventitious, what fundamental and what secondary." Frazer did build conceptual castles in the air, to be sure; but he built them out of the odd facts of human life and history—their "imponderabilia"—which he loved better than any abstraction.[117] It was that ability to draw large generalities from small particulars that in Malinowski's view made Frazer not just a great imaginative artist but also a "pioneer in modern scientific anthropology."[118] Robert Thornton and Peter Skalnik make a similar point about the Ethnographer himself: that his "substantive metaphysics" was what distinguished his interpretive ethnography from the lifeless factual inventories of prewar ethnology.[119] It also distinguishes the empirical theorizing of Malinowski's Trobriand monographs from his later theorizing about culture in general. Like his Trobriand coresidents, Malinowski reasoned more comfortably with particulars than with abstractions, and to better effect in the living contexts of fieldwork than in an office. Like his early role model James Frazer, he was a teller of stories—Tolilibogwa, the man of songs. But unlike Frazer's, the stories he told were firsthand *inside* stories.

CHAPTER 3

HOBO SOCIOLOGIST

NELS ANDERSON

It is a curious fact that though British anthropologists invented the practice of participant observing, it was an American social philosopher and educator, Eduard Lindeman, who introduced the term "participant observer" into public discourse, in 1924. What he meant by that term, however, was not what ethnographers would later mean by it. Lindeman's "participant observers" were not sociologists, but their subjects: "participants not in the sociological study but in the group being observed." Their role was "to provide scientists with facts, criticize categories and discover new ones, [and] correct any conclusions that reflected scientists' views rather than the group's." Subjects could conceivably be both resident and expert, Lindeman acknowledged, but it was highly unlikely, and not desirable, since residents trained to observe as scientists would thereby lose their value as untainted witnesses. Lindeman did not mention the possibility of training sociologists to reside and observe. His conception of science was thus conventionally dichotomous: living in society and observing it scientifically were distinct and mutually exclusive activities.[1]

However, American sociologists in the 1920s and 1930s *did* practice some kind of resident or participant observing: not the Malinowskian kind, but their own homegrown varieties, which grew out of indigenous traditions of empirical sociology and the related practices of social work and social survey. Nels Anderson's study of hobos in the early 1920s—one of the first field projects of the Chicago school—was unambiguously resident and participant, though the resulting book could never be mistaken for Malinowskian ethnography. Anderson's field practice was distinctly his own, drawn less from any scientific model (he knew little of those) than from his own life as a hobo; and the project was designed to serve the practical end of aiding homeless men. Another clear case of participant observing by a Chicago sociologist is Albert Blumenthal's dissertation study of his Montana hometown, "Mineville." Others in the 1920s and 1930s also edged toward more fully situated practice without actually residing.

The perception of the Chicago school as a nursery of participant observing has been challenged by the historian Jennifer Platt. The term "participant observation" was entirely absent from sociology textbooks before 1930, she found, and even when it did gain currency, in the 1930s and 1940s, it was loosely applied to a range of practices that included observing by actual residents and fieldwork by sociologists to follow up on formal interviews. Platt concluded that none of the sociological field studies from the interwar decades qualify as participant observation in the true—that is, the ethnographic—sense. It was not until the 1950s, Platt argues, that participant observation was "effectively invented, for wide diffusion," by a new generation of Chicago sociologists—Everett Hughes, Anselm Strauss, Howard Becker, Erving Goffman, and others—who practiced an ethnographic sociology of small groups.[2]

Platt's historical aim was to make a fuzzy category of sociological method precise and rigorous by restricting it to mean exclusively what anthropologists did. In that narrow sense it is true enough that sociologists were not ethnographers before the 1950s. But restricting and purifying the category in this way obscures a more messy historical reality of impure practices. Sociologists in the 1920s and 1930s devised their own distinctive methods of inside science to fit their particular subjects and aims. These indigenous practices included study by sociologists of their hometowns, social or sociological surveys that sought to understand social problems by using the methods of empirical science. This diverse historical reality is more faithfully captured by a fuzzy analytical category than by one that strives for precision and rigor. It is not precise definition that matters, after all, but social process. How and why did sociologists in Chicago and elsewhere first take to situated work in the field? And how and why did a few of these field studies turn into sociological versions of resident or participant observing, while most remained within established modes? Were there practices in sociology that were analogous to intensive ethnological survey in social anthropology, and that created situations in which a more fully resident science might develop? To address these questions I will need to go more deeply into the history of the social sciences than was necessary in the case of ethnography.

SOCIAL SURVEY AND HOMETOWN SOCIOLOGY

Sociology in the United States in the late nineteenth century was not a science of the field. Like British ethnology it was a kind of social philosophy, whose practitioners sought to discover universal laws of human social development in history, working mainly in libraries, with texts. Sociologists also shared with ethnologists a growing conviction that progress in their science would come from improved methods of empirical investigation, especially firsthand gathering of facts in situ.[3] The founders and heads of the leading departments of sociology, Franklin Giddings (Columbia University) and Albion Small (University of Chicago), were both ardent advocates of hands-on research.[4] As James Williams, a Giddings disciple, put it, "There will be no such thing as sociology until we have begun at the A, B, C of method—observation. . . . What sociology most needs is *field-work*."[5] Leading textbooks of the 1890s, including Small's, enjoined sociologists to go afield.[6]

What that would mean operationally, however, was not so clear. Whereas social anthropologists were drawn to far-off places that to Europeans were strange and exotic, sociologists worked in communities that were near to hand and generally familiar. So while anthropologists had no scientific precursors and competitors in fieldwork, sociologists decidedly did. Three occupations—philanthropic social survey, professional social work, and investigative journalism—were already established, pursuing practical ends empirically in situ. For sociologists modern societies were thus contested subjects, and devising appropriate ends and means of empirical study was consequently more complicated for them than for ethnographers, who had to contend only with amateur travelers and missionaries. Sociologists had more to borrow, contest, or disown.

Of the three established traditions of social field investigation, the most developed was the social survey. From its late nineteenth-century origins in Britain (Charles Booth, Seebohm Rowntree, Sidney and Beatrice Webb) and the United States (Jacob Riis, Jane Addams, W. E. B. Du Bois), social survey in the 1910s and 1920s became a veritable industry, especially in the United States, where the movement was organized, funded, and promoted by the Russell Sage Foundation and other organizations.[7] By 1928 no fewer than 2,775 surveys had been carried out by an array of church, civic, charitable, and academic organizations. Although the vast majority of these (2,621) were small and local, and focused on some special issue—education, health, industrial labor conditions, urban planning, delinquency and corrections, and housing were the top favorites—a few (154) were comprehensive surveys of entire communities. These delved into more sociological issues of ethnicity, family and religious life, class and social mobility, and social attitudes.[8] The purpose of social survey was philanthropic—to improve society through organized intervention. But its means were devotedly empirical. The idea was that interventions should be guided by knowledge of the facts, not just charity or sentiment. So a thorough survey of the facts on the ground was the essential first step in any project.

In social surveys, facts were gathered opportunistically from any and all sources: archived records of courts, schools, welfare agencies, and charity organizations, as well as census-style door-to-door interviews of neighborhoods, using standard questionnaires. These methods produced numerical data in standard categories that could be quantified and tabulated for use in designing policies and remedial projects. Survey organizers valued community initiative and depended on local volunteers for canvassing and consciousness raising. Surveys were variously run by volunteers and (increasingly) professionals in social work and charity organizations, as well as by political activists and, more rarely, sociologists.[9] For example, W. E. B. Du Bois was a trained social scientist and carried out his one-man, thirteen-month social survey of the black population of Philadelphia's Seventh Ward (1897–98) to demonstrate scientifically that the deep causes of racial prejudice were social and economic.[10] As one of the founders of the survey movement put it (approvingly), survey was not scientific research alone, nor charity, nor civic uplift; but all of these rolled into one.[11]

The scale and logistics of social survey ruled out sustained contact with residents, as they did in the case of ethnological survey. But other social services did bring investigators into more intimate contact with subjects on their own home ground. As social work became a university-based profession, home visiting and family casework became its defining practices.[12] The settlement-house movement that spread from Britain to the United States in the late nineteenth century was another such activity. Designed to give resident social workers and college students access to working-class urban neighborhoods, settlement houses became in situ outposts for local surveys and remedial interventions. Jane Addams's Chicago Hull-House was the paradigmatic example.[13]

Finally, there was investigative journalism, which was in its heyday in the prewar era of muckraking and progressive reform. Newspaper reporters had always worked the streets, of course; but investigative journalists went deeper, going undercover to work incognito in factories or fields or to live in lower-class neighborhoods to get the behind-the-scenes, inside stories of life in industrial and capitalist mass society.[14] Investigators were not scientists. They pursued literary and propaganda ends, and few were undercover for more than a few weeks or months. However, they produced facts that were more revealing of the everyday lives of ordinary people than those of social survey, and more like the facts that would one day constitute resident sociology. Exposés were not scientific reports; however, vivid stories of actual lives could inspire curious and idealistic young people to take up social science.

In academic sociology the impetus to empirical field study was not practical or philanthropic but pedagogical. It came from the widespread movement in colleges and universities in the late nineteenth century to make learning less passive and bookish and more actively hands-on, by enabling students to participate in knowledge making. In the natural sciences, campus or seaside laboratories afforded dedicated space and equipment for student practicums; in the social sciences, getting students out of libraries and into the world was less straightforward. Without labs or tools, sociologists were obliged to improvise makeshifts for situated study, using whatever resources were ready to hand.

One of these makeshifts was to get students to draw on their own life experiences for empirical material. Student practitioners of this "hometown sociology" were in a sense their own native informants. It was not actual fieldwork, since it relied entirely on memory, but a kind of virtual fieldwork—a science in recollection. And it had the great advantage of being doable right away without new resources or infrastructure: students only had to see their lives in a new way. The Chicagoans Albion Small and George Vincent in their influential 1894 textbook conjured up an "anonymous but not fictitious Western settlement" as it evolved from a pioneer crossroads on a prairie river to a modern industrial city, complete with imagined maps of its stages of growth and details of its changing social order. The exercise was imagined, but not imaginary. Small and Vincent's nameless town was most likely Vincent's hometown of Rockford, Illinois, and its use in a sociology text was meant to inspire students to think sociologically about their own hometowns. The point was that sociology was not abstract social philosophy but a living subject that students had experienced in their own lives, without knowing it as sociology. Their lives were data for their own scientific inquiries.[15]

Judging from the roster of prewar Chicago dissertations, it appears that no students took the bait set out by Small and Vincent.[16] Social philosophy remained the favored, high-end mode of sociology in elite universities. However, some Chicago graduates did later get their own students to engage in classroom exercises in hometown sociology. G. P. Wycoff, who had studied with Vincent, assigned this exercise to his students at Grinnell College in the late 1890s. And Jesse Steiner (PhD 1913), as a professor at Tulane in the early 1920s, had students in his graduate seminar write hometown sociologies. Hundreds of these virtual field projects were carried out; and though most were short exercises, they are a sign that interest in situated sociology was alive at the collegiate grassroots in the postwar years.[17]

For the first sociologists to attempt actual rather than virtual fieldwork, social survey was the most usable model. Although social survey was not sociology, it was empirical investigation; and since the two occupations were often conflated in the public mind anyway, survey was an available resource that could be turned to sociological ends. As one early advocate of fieldwork put it, social survey was "one of the best answers to the assertion sometimes made by ignorant or prejudiced people that sociology is not and cannot be scientific." If teachers lacked the time for fieldwork, their students constituted a pool of available (and free) labor for projects that combined research and instruction.[18] A few professors of sociology did organize social surveys, mainly in state colleges and universities, whose institutional mission of public service made them supportive contexts for such ventures.[19] Social surveys were for these sociologists what state experiment stations and extension services were for scientists in schools of agriculture. In urban universities, settlement houses in poor neighborhoods afforded similar ready-made opportunities for participation in organized social survey.

The appeal of direct participation in social surveys faded, however, as sociologists sought to raise their standing among the sciences by developing scientific practices of their own, and by leaving social interventions to the service occupations and professions. For a time sociologists looked to the vast accumulation of survey data as a windfall resource that they could repurpose to scientific ends, thereby saving themselves the labor and expense of actual fieldwork. They also saw a role for themselves in providing departments of social work with basic instruction in science and guidelines to proper scientific methodology.[20] But in the 1920s sociology and social work went their separate ways. Social workers did not relish sociologists dictating their standards of practice, while sociologists found that social-survey data were less amenable to repurposing than they expected: too generic and preprocessed, and too much concerned with social pathologies and too little with normal activities of everyday life. As the historian Martin Bulmer put it, sociologists regarded social survey methods as "tainted" by the interventionist aims they served.[21] Sociologists thus learned what anthropologists and others had learned before them: they could not delegate empirical work but would have to collect data themselves in situ by their own means and to their own scientific ends.

In fact, social survey also disappeared from social work in the 1920s, as social workers looked to clinical psychology and psychiatry as the prestigious theoretical disciplines that would give their profession academic standing. Sociologists, meanwhile, took up a kind of community study untainted by reform agendas and propaganda. Bulmer describes this divergence as a total discontinuity.[22] Yet elements of social survey did persist in sociology. Community study was a kind of survey: not social, now, but sociological. And elements of hometown sociology likewise survived. As sociologists took to actual fieldwork, hometowns were not infrequently the communities they elected to study. Social survey and hometown sociology were the vehicles that took sociologists out of libraries and into firsthand investigations in situ. This pattern is apparent in the first prewar ventures in fieldwork by sociologists at Columbia University and the University of Chicago.

In New York at least six inside studies were carried out as dissertation projects: four of small rural towns, two of urban neighborhoods.[23] Of the small-town studies three at least were carried out by former or current residents. James Williams had lived for several years in "Blankville," an isolated rural village somewhere in the eastern United States, whose residents he knew as friends or acquaintances. Newell Sims had likewise been a longtime resident of "Aton," a village in the Indiana lake district, which he realized would be a fine subject for empirical study (along with a nearby town, "Bton"). Warren Wilson chose for his dissertation subject a small Quaker enclave in the Hudson Valley, "Quaker Hill," where he had lived since 1893.[24] Prior knowledge of their subjects thus seemed a necessary prerequisite to sustained fieldwork, and many college students at the time had been born and raised in small towns.

With urban neighborhoods, prior experience likewise seemed a prerequisite to fieldwork, though not a resident's experience. Few collegians would have known urban slums as residents; if they did know them, it was likely to be from novels or exposé journalism, or by taking part in some kind of reform activity. The two Columbia studies of urban communities were carried out by students with experience of social survey. Howard Woolston's study of Manhattanville, a polyethnic community at the far northern end of Manhattan Island, began as a fact-finding survey by the occupants of the Speyer School, a progressive school and settlement house, in preparation for philanthropic programs in the neighborhood. It then evolved into a scientific investigation of the social dynamics of urbanization in what had until quite recently been a rural village but was rapidly becoming a neighborhood of the great metropolis. Woolston had lived in Manhattanville for two years (most likely as a resident of Speyer House) and hoped that inside study of the locale would reveal "the forces that are moulding the city."[25]

At the other end of Manhattan Island, Thomas Jones ventured into another former "village"—Greenwich Village—to study an ethnically diverse working-class enclave that, like Manhattanville, was experiencing a disruptive invasion by exotic outsiders: not ethnic immigrants, in this case, but artists and bohemian refugees from bourgeois society. To probe the dynamics of this urban transformation, Jones drew on his own varied experience in social services: as public-school teacher, acting head worker of Columbia's settlement house, canvasser for the Charity Organization Society and the Federation of Churches, and agent of the US Census Bureau. Though Jones was not a Village resident, he knew all the tricks of getting through front doors and into family parlors, and of persuading strangers to answer intimate questions about themselves and their neighbors. Every Saturday he made twenty such visits, knocking on doors and announcing a "sociological census"—with emphasis on the "census," lest the unfamiliar "sociology" cause wary residents to slam their doors in his face. Visits were necessarily brief, but Jones took advantage of them to discreetly record informants' manners and attitudes.[26] The benefits of inside observing were soon apparent. Jones was struck by the marked difference between the information he got as a home-visiting sociologist and what he had gathered as a church or settlement canvasser. As an agent of philanthropic institutions, he had access only to the self-selecting minority of residents who saw these institutions as a way out of the tenements and into the middle class. As a sociologist, he had access to a cross section of a diverse urban society and to community life as residents lived it.[27] Philanthropic survey was thus serendipitously repurposed to sociological ends.

At the University of Chicago the strong traditions of social survey and social work were the chief resources for sociologists looking to work in situ. There was the famous trio of Jane Addams, Sophonisba P. Breckinridge, and Edith Abbott (who had a part-time appointment in sociology from 1913 to 1920), as well as Charles Henderson, a German-trained clergyman, sociologist, and social activist. As professor of ecclesiastic sociology and university chaplain, Henderson taught courses in both sociology and social survey and organized field investigations at the university's settlement house in Chicago's industrial South Side.[28] Unlike Small and Vincent, Henderson had the firsthand knowledge and institutional means to get students into working-class neighborhoods. Charles Bushnell (PhD 1902) surveyed stockyard workers, both in their workplace and in their homes and neighborhoods. John Gillette (PhD 1901) carried out a similar study of steelworkers in the giant mills along the Calumet River.[29] Although these studies had a reformist aim, they were also meant to advance Henderson's view that problems of industrial labor could only be understood sociologically, in the context of workers' family and community relations. So in addition to the generic methods of census and questionnaire, Bushnell also observed street life and visited and interviewed workers and their families at home.[30] Looking back in 1928 at these first efforts, the sociologist Luther Bernard saw them as "the beginning of the use of the [social] survey for purposes of sociological generalization."[31] Lacking tried-and-true field practices of their own, sociologists used their experiences as hometown kids or social-service snoops.

THE CHICAGO SCHOOL

Situated investigation in sociology between the wars was most systematically and fruitfully pursued in what came to be known as the "Chicago school."[32] Students were pushed to work in situ, and personal engagement with subjects was a feature of many student projects, extending in several cases to actual residence. This outpouring of original work was the result of a program and an organization that went well beyond the makeshifts and piggybacking of hometown sociology and social survey.

The Chicago department was transformed in the late 1910s by a near-total turnover in personnel: the death of Charles Henderson in 1915; the growing influence of W. I. Thomas, who was assembling documentary material for *The Polish Peasant*; and the appointment of Robert Park as professorial lecturer in 1913, followed by Ernest Burgess in 1916 and the social psychologist Ellsworth Faris in 1919. Thomas's dismissal in 1918 on trumped-up morals charges left Park as the department's guiding intellectual force. And the exodus of the department's social workers into a school of their own, in 1920, left the sociologists in sole possession, most notably of the practical course in field method that Edith Abbott had long taught, more as social survey than as sociology.[33] In the early 1920s a fruitful collaboration married Park's vision of an empirical, situated urban sociology to Burgess's practical devotion to organization and methodology.[34] The reconstituted department's communal aim, Herbert Blumer recalled, was "to catch and respect human group life as it is being lived, to recognize that it is embodied in the experience of its participants and that it must be dug out of that experience, . . . to enter into the ongoing collective experience of people."[35]

This vision of an in-the-world, in situ sociology derived less from prewar academic roots than from the personalities and life experience of its two founders. Thomas and Park (fifty-one and fifty-two years of age, respectively, in 1915), though highly educated, were first and foremost men of the world: energetic and charismatic personalities insatiably curious about people of all sorts from ordinary to outlandish, and drawn especially to the marginal and deviant. Thomas pursued such diverse intellectual subjects as folk psychology, sexuality, race, and ethnicity, and took up an unorthodox bohemian lifestyle.[36] He had the sensibility of a "literary man in the reportorial sense," Park recalled. "He wanted to see, to know, and to report, disinterestedly and without respect to anyone's policies or program, the world of men and things as he experienced it."[37]

That was an apt description of Park himself. Following formal study in the United States and Germany, he worked for ten years as a city reporter and editor (1887–98) and for another ten (1903–13) as an investigator and publicist, first for the Congo Reform Association and then for Booker T. Washington, at Tuskegee, where Thomas had discovered Park and persuaded him to take part in creating an empirical science of sociology at Chicago.[38] Their plan was to bring in men of worldly experience as part-time faculty, who would expose students to the lived realities of social life. The knowledge that Park most valued was knowledge gained in an active life: "the personal and individual knowledge which makes each of us at home in the world in which he elects or is condemned to live."[39] Whatever engaged the interest of newsmen and their readers was for Park also the proper subject matter of sociology. "One might fairly say," he wrote, "that a sociologist is merely a more accurate, responsible, and scientific reporter." He was drawn especially to the inchoate life of burgeoning megacities like Chicago, where new social types and new ways of life were in the making and could be observed inside and up close.[40]

Park's sociology resembled the hometown sociology of Small and Vincent in its emphasis on personal experience. But it was actual, not virtual experience, and the urban society of mass immigration and explosive growth that captivated Park was a world apart from the socially homogeneous small towns to which Small and Vincent had looked for subjects.[41] Park exposed students to forms of social life unlike anything they would have known in their own lives. His worldly empiricism also resembled the socially engaged empiricism of Chicago's reformers and social workers, but again with a difference. Park respected Jane Addams and Edith Abbott and took an active part in two major social surveys of race relations in Chicago and in the Pacific coast states.[42] However, he regarded social survey and sociology as distinct activities that should be kept carefully separate. "Park was not a reformer by nature," his biographer wrote. "His passion was for observing rather than changing." Students eager to turn their sociology to social causes were bluntly set straight: the world was full of crusaders; their task as sociologists was to investigate human society in the same disinterested and objective way that zoologists sought to understand the life of insects.[43]

The investigative soul of Chicago sociology was thus the engaged and omnivorous interest of the newsman, the curious urbanite and connoisseur of human oddities, the sociological naturalist.[44] This situating empiricism informed Park's landmark 1915 essay "The City," which laid out a capacious program of urban sociology, and the 1921 textbook by Park and Burgess became the operating manual of the Chicago school.[45]

The geography of interwar field studies was also markedly different from that of prewar surveys. Whereas Henderson had concentrated on South Chicago and issues of industrial labor, Park and Burgess focused on Chicago's Near North, a culturally and economically diverse area just north and west of Chicago's downtown Loop, and an area of ever-shifting socioeconomic zones. It was an ideal place in which to observe and study a wide range of human social behaviors in context. A sociological "laboratory," Park and Burgess called it.[46] But given their devotion to concepts of ecology, it is more apt to think of the Near North as a human ecosystem roiled by invasions, retreats, displacements, and successions—a human natural history. In fact, as Thomas Gieryn has shown, Chicagoans relied equally on analogies with lab and field in their programmatic characterizations of an urban sociology.[47]

Of the 209 student dissertations produced between 1918 and the late 1930s (76 PhDs, 133 MAs), over two dozen were carried out mainly or partly in situ.[48] The subjects of these projects assort into several groups. The main group of thirteen were studies of communities or "natural areas" and their distinctive inhabitants: "Hobohemia" and its itinerant hobo laborers (Nels Anderson, 1923), the lakeshore Gold Coast and adjacent Little Italy (Harvey W. Zorbaugh, 1929), and the area of taxi-dance halls (Paul Cressey, 1930) are the best known of this group. A second group of nine took as their subjects occupational types not confined to a single natural area: youth gangs (Frederic Thrasher, 1926), school truants and juvenile delinquents (Clifford Shaw, 1929), and "jack-rollers" specializing in robbing drunks (Clifford Shaw, 1930) are exemplary. And a third group of six took Chicago methods to communities farther afield, including a small mining town in Montana (Albert Blumenthal, 1932), a tight-knit community of immigrant Swedes struggling with assimilation in Austin, Texas (Carl Rosenquist, 1931), and an insular colony of Russian pietists marooned in Los Angeles (Pauline Young, 1932).

It was not just personal interests that produced this diverse harvest, but the social machinery created in the early 1920s to systematically get students into the field and guide them in their own research projects.[49] A methods seminar with a field practicum introduced students to empirical methods, and field projects were a routine feature of advanced seminars. It was not enough to dirty their hands in dusty archives, Park liked to tell students; they had also to get the seats of their pants dirty sitting in hotel lounges and on flophouse steps, in slum shakedowns and burlesque and orchestra halls.[50] He would take them on walking tours of neighborhoods, introducing them to people and getting them used to walking, talking, and noting all that was going on around them. Students who showed a taste for fieldwork were eased into thesis projects, with grants-in-aid from the Laura Spelman Rockefeller Memorial.[51] Park recalled that his experience as city editor shepherding a staff of beat reporters served him well when it fell his lot to tend a flock of student field-workers: in Martin Bulmer's words, "pushing, suggesting, inquiring, rewriting, needling, scolding . . . in order to fulfill the plan of studying the city in all its aspects."[52] The prewar generation of senior sociologists had dreamed of a system in which they were paid to do research, with students assisting. Park and Burgess reversed that order: their first priority was to get students working in the field with faculty guidance.[53] Scientific production and social reproduction were thus combined in a fruitful symbiosis.

Infrastructure was also created to conjoin individuals' projects into one large communal one. This was largely the achievement of Vivien Palmer, who was appointed senior research worker in 1925 to oversee and coordinate students' researches. Palmer completed Burgess's project of defining and mapping the human ecology of Chicago's seventy-five "natural areas," adding cultural and historical criteria to Burgess's geographical and institutional ones. She assembled a complete areal map of the city and had copies run off for everyone, and compiled the accumulated data of individual projects into a comprehensive Local Community Fact Book. With a foundation grant she set up a calculating room with machines for numerical and statistical data processing, and a map room where the accumulated knowledge of natural areas and socioecological zones could be easily consulted. And her handbook of field methods brought together the group's working methods and collective experience.[54] Each student project became part of what was in effect a grand sociological survey of Chicago by many hands. Howard Becker (PhD 1951) likened it to a mosaic, each new piece adding to an overall picture that became more clear with each addition. Beginning his own research a few years later in San Francisco, Becker automatically looked for a fact book of his new community, only to discover that what he had taken for granted in Chicago existed nowhere else.[55] Organization thus went a good way to removing the impediments to situated research that had till then kept most students in the library.

Just how deeply situated Chicago sociology was has been disputed. Martin Bulmer identified some projects as "pioneering works of participant observation," and that view was casually absorbed by the potted histories of methodological handbooks.[56] This view was challenged in the 1980s by Jennifer Platt and others, who argued that fieldwork was not the dominant practice of the Chicago school but just one among many, mostly documentary; and that what had been called "participant observing" was no more than traditional social survey and case study.[57] It is true that Chicago sociologists rarely if ever practiced the pure participant observing of ethnographers; all commentators have agreed that eclecticism was a hallmark of the Chicago school.[58] Yet intensive empirical work in natural situations was no less its hallmark, and several projects were clear instances of resident science. So it is not surprising that the revisionists were in their turn revised by alumni of the Chicago school and historians, who sought a more nuanced middle view. The Chicago graduate Ruth Cavan remembered Parkian street practice approvingly as "neither case studies nor statistical surveys . . . but rather observation, participant and otherwise." The historian Mary Jo Deegan documented a distinctive tradition of "qualitative" research, which she labeled "ethnography."[59] And Shane Blackman wrote of an "ethnographic mosaic" of qualitative strategies that included interviewing, observing, participant observing, and life histories.[60] Platt herself edged toward that middle view when she wrote that "what they [Chicagoans] were doing included observation as a participant, [but] it was not 'participant observation.'"[61]

Chicago school sociology may be most aptly likened not to Malinowski's participant observing, but to the intensive ethnological survey out of which that practice unexpectedly evolved. Like the survey ethnology of Seligman and Haddon, the sociology of Park and Burgess was a situating practice with the potential to become, in favoring circumstances, resident or participant observing. In both cases intensive local work created opportunities for deeper resident study, while creating structured impediments to such work. The question is, What particular features of Chicago field practice enabled, and what ones limited, a resident practice?

SOCIOLOGICAL SURVEY

"Sociological survey" was a new term in the 1920s. Vivien Palmer was the first to define it, in her guide to field practice, though it's not clear that she minted it. With its play on "social survey" Palmer's neologism staked sociologists' claim to a domain of empirical investigation to which social workers had a prior claim. It marked a boundary between two professions that had been uneasily allied but were then drawing apart: the one a practical profession, the other an inductive science.[62] Palmer warned sociologists firmly against reusing the data of social survey to their own scientific ends. However tempting, that shortcut could only impede the development of properly sociological methods; it was, Palmer asserted, "one of the stagnant backwaters in sociological research."[63]

Interview was the dominant practice of sociological survey, as it had been for survey ethnologists, and it was the subject of much methodological cogitation. Palmer distinguished various subtypes of interview: formal, informal, initial (to break the ice), and repeated (to record entire life histories); with strangers, friends, acquaintances. She laid out the best ways of making contact and establishing rapport, avoiding observer bias, balancing spontaneity and control, and discreetly noting subjects' body language and demeanor.[64] Interview by standard questionnaire was the easiest way to gather generic facts, but it lacked the personal perceptions of informal conversation. How actively interviewers should intervene was uncertain. Was undirected listening more likely to elicit true facts, or was active probing and questioning? Burgess likened formal interview, approvingly, to cross-examination in a court of law, in which questioners could draw out hidden facts and guide or correct witnesses. Active questioning seemed to him more scientific, more like controlled experiment, than passive listening. On the other hand, in free-flowing conversation interviewers could enter into their subjects' mental and emotional lives, especially if interviews were recorded verbatim. "To record the interview in the words of the person signifies a revolutionary change," Burgess wrote. "It is a change from the interview conceived in legal terms to the interview as an opportunity to participate in the life history of the person, in his memories, in his hopes, in his attitudes, in his own plans, in his philosophy of life."[65] Students were urged to get into neighborhoods and talk to people in ordinary situations and in their own vernaculars.[66]

The most engaged and intimate kind of interviewing was taking down a subject's full life history in repeated interviews. The case method or case study method, Chicagoans called it, and it was famously exemplified by Clifford Shaw's collaboration with Stanley, the teenaged "jack-roller." Shaw worked actively with Stanley to produce his fifty thousand–word life history: compiling a record of Stanley's numerous misdemeanors, delinquencies, arrests, and incarcerations as an aide-mémoire; checking his facts and pressing him to correct or elaborate.[67] Intensive life histories enabled interviewers to get at the sentiments and feelings that motivate social behavior, yet go largely unexamined in more formal interview.[68] It also enabled them to put individuals in the larger social context of family and communal life—the "total situation," Frederic Thrasher aptly termed it.[69]

Life-history interviewing was a deeply situating practice. Indeed, its advocates sometimes make it sound very like participant observing. As Charles Cooley wrote, "We aim to see human life as an actual dramatic activity, and to participate also in those mental processes which are a part of human function and are accessible to sympathetic observation by the aid of gesture and language."[70] And Ernest Burgess: "[T]he social worker aims first of all to put himself . . . in the place of the other person, to participate in his experiences, to see life . . . as the other person sees it."[71] Life-history interview was thus a kind of affective, psychosocial participation in subjects' inner lives: not the actual participation of coresidence, but a step in that direction.

Observing, unlike interview, was for Chicagoans not a subject of self-conscious reflection. Park and Burgess enjoined students to always listen and observe, but they had no developed methodology of engaged observing. Herbert Blumer recalled that it was seen "as being a contributory device, and for that very reason was not studied carefully."[72] Vivien Palmer's treatment of observing in her field guide was perfunctory, consisting mostly of commonplace reflections on subjective bias (the so-called personal equation) and the uncertain value of appearances as evidence.[73] Palmer thus took a narrow view of observing as passive onlooking at ritualized events—meetings, parades, celebrations—where outside watchers could only guess at what these meant to participants.[74] The great advantage of the social sciences, Palmer asserted, lay precisely in its "new modes" of interview and introspection—that is, personal testimony—because these produced direct evidence of subjects' views and feelings. The natural sciences were in contrast "limited" to detached experiment and observing—a bravado inversion of the orthodox hierarchy of the sciences![75] If it was active interview that gave sociology its identity and authority as a science, as Palmer suggested, it is no wonder that Chicagoans gave observing little thought and a low epistemic rating.

Many sociologists at the time seemed to doubt that actual residence was routinely feasible in fieldwork. Palmer did implicitly acknowledge that it might be when she distinguished two subtypes of "participant observer" within Eduard Lindeman's definition: the resident trained to observe scientifically, and "a person who has identified himself with a group simply for the purpose of studying it"—that is, a resident sociologist. She went on to say that closed "interest group" communities—gangs, clubs, religious sects—were accessible only to an "adopted insider" sociologist. However, she thought that sociologists would learn more in groups in which they were already known and accepted than in those in which they were strangers—a decided limitation.[76] Her concept of resident sociology was more or less that of hometown sociology.

In fact, Chicago sociologists were quite good at connecting with informants. However, most used their access not to take part and observe but to set up formal interviews. Residing was a preliminary to science, not science itself. Frederic Thrasher's study of youth gangs exemplifies this pattern. Gangs were a paradigmatic "interest" community. Exclusive and suspicious of outsiders, gang members could not be directly approached; outsiders had somehow to become adopted insiders. Thrasher's opening presented itself one day when, driving in his car, he drew up behind some boys joyriding on the back of a delivery truck. He began a racing game, speeding up as if to crash into them, then backing off. As Thrasher expected, the boys begged for a ride; and he obliged, as recklessly as he dared. He then asked the boys if they would like to form a Sunday athletic club at his apartment, which they did, and after that a handball team at a settlement-house gym. The group was soon, as Thrasher put it, "ripe for study along any line." Boys began asking his advice on family and personal matters and were easily drawn into telling him the particulars of their street lives. As the club's "secretary," Thrasher could record, in the guise of taking minutes, what were in fact free-flowing sociological interviews.[77] This was not resident observing: Thrasher's boys were more participants in his project than he was in their street life. He participated only to set up interviews, and these were carried out not in the boys' own social spaces but in a place created for the purpose of science—a kind of makeshift field lab or settlement house.[78]

Paul Cressey employed similar stratagems to secure informants about the practices and experience of taxi-dance halls, where lonely men bought brief, cheap, and anonymous companionship. The best way to set up interviews, Cressey learned, was for field-experienced students to pose as customers, dance with girls a few times and buy them drinks, and maybe see them home; then persuade them to relate their life stories. (Dancers' parents were easier: Cressey's semiofficial status as an employee of the Juvenile Protective Association was an acceptable reason to ask personal questions.) Interviewing customers was trickier. Cressey learned how to pose as an ordinary guy and enter into casually anonymous and often intimate talk about guy stuff. However, the slightest slip in his cover persona abruptly ended the relation. It was not a practice for beginners. Martin Bulmer referred to Cressey's practice as ethnography, but it wasn't quite that. Cressey called it "case study," but it was more than that.[79] It was in hindsight a practice in transition from case work to resident observing. Typically projects stopped short with interviewing and life histories, as did Thrasher's, Cressey's, and others'. Kimball Young, in his study of a "disintegrating neighborhood," conversed casually as himself with people where they lived: on front stoops, in grocery stores, everywhere but in saloons, where well-dressed strangers who talked like dictionaries (as he put it) were decidedly unwelcome. He would have done better, he realized later, had he lived in the neighborhood and dressed and acted like a neighbor.[80] The opportunity was there, unrecognized.

It may be that the social machinery for getting novices routinely into the field inadvertently hindered them from pursuing opportunities for resident study. The routines of course work and teaching made it hard logistically for students to live among their subjects and favored a practice of repeated visits—a *commuting* science, so to speak, not a *residential* one. The strategy of using Chicago's Near North as a handy sociological laboratory created additional incentives to commute. With research sites so quickly and cheaply accessible by public transport, who would choose to reside? And the abundant data ready to hand in the Local Community

Fact Book, map room, and library made it less necessary to gather new facts in situ. According to Kimball Young, Harvey Zorbaugh lifted much of his slum data from Young's own dissertation and most of his Gold Coast data from the Chicago social register, not from actual work in clubs and drawing rooms.[81] It is even possible that the communal "mosaic" of sociological survey nudged students to frame their projects as surveys, to contribute to the communal endeavor. So it is perhaps not surprising that commuting, interview, and neighborhood survey became the default practices, and that few students took the opportunities that arose to reside, participate, and observe.

It is thus no accident that the most fully resident field projects were located outside Chicago, in places where there were practical incentives to live among one's subjects, and none of the campus amenities and shortcuts that favored a commuting science. The subjects of these projects tended to be closed communities—small towns, insular ethnic or religious groups—of the sort that Vivien Palmer singled out as fit subjects for "adopted insiders." And for strangers from afar, residence and participation would be the natural, even the only, way to be there. Distance favored a resident over a commuting practice, much as it did for anthropologists far afield.

Pauline Young's five-year study of an insular Russian religious community in Los Angeles illustrates this pattern. At first Young pursued her project as a Chicago-style sociological survey, but realized that she could make no sense of her accumulating data without deeper knowledge of her subjects' inner lives and feelings—knowledge that she would get only by taking an active part in the life of "Russian Town." As Robert Park put it, she would have to "penetrate the inner sanctum of Molokans" and "think and feel Molokan." And so she did; so well, in fact, that Park had repeatedly to warn her against allowing emotional involvement with her subjects to override scientific detachment.[82] Young, it appears, became a participant observer almost to the point of going native. Her insular and unmodern subjects, together with her own separation from her academic home base in Chicago, allowed or even called for engaged residence.

A more clear-cut (and better-documented) case of participant observing was Albert Blumenthal's community study of "Mineville"—the cover name he gave to his hometown of Philipsburg, Montana. Ernest Burgess described Blumenthal's project as "hometown sociology" and his role as "participant observer" in Lindeman's meaning of a resident who also has the competence of a scientific observer.[83] Blumenthal's project was the older, makeshift practice brought up to date by the improved methods of the Chicago school. Blumenthal lived among his former neighbors in Mineville (population 1,410) for two and a half years, participating in "the variety of experiences that make up town life" and engaging in "friendly conversation in which the other person communicates his experiences, feelings, and attitudes much as if he were talking to himself." Recorded more or less verbatim, these conversation interviews were his main source of field data. Generic questionnaire data—the usual stock in trade of community survey—were a minor complement, in part because these were hard to gather discreetly in a small town. A house-to-house census was quickly abandoned when it threatened to ignite "the running fire of gossip." Mapping was abandoned for the same reason. (Aerial mapping would have been more discreet and accurate, Blumenthal thought).[84] Blumenthal's subject and his hometown situation thus made residence and engagement the path of least resistance and ruled out survey as a fallback.

Blumenthal may in fact have come to Mineville intending to carry out a standard community survey. His cover story was that he was researching a history of the town and its institutions, and the story may have been true. But as conversational interviews of residents took him into unexpected depths of personal feelings and experience, he realized that these were opening up a kind of sociology quite different from the survey mode. As he wrote to Burgess, "It seems to me that the most promising results of my study is [*sic*] that of breaking the ground for the development of a technique of making intimate studies of small communities." Conversation interviews became a kind of participation in individuals' lives and, as Blumenthal put it, "the most concrete revelation I have ever encountered, and only my peculiar role as confidant could have enabled me to secure it." It was the personal engagement with subjects that afforded insights into how small-town society worked, not detached objectivity, "the common fallacy of sociologists."[85]

Blumenthal's experience thus recapitulated Malinowski's in Omarakana, when conversing and observing as a resident unexpectedly punctured the ethnographer's positivist faith in his "method of objective documentation" and revealed a better way into the deep operating principles of Trobriand society. Blumenthal's book, *Small-Town Stuff* (1932), reinforces the parallel: like Malinowski's monographs, it focuses on the institutionalized activities that give a community its distinctive structure and mores. In Blumenthal's book that activity is gossip—the tyranny of gossip and conformity that was at the time a stock literary trope of novels and exposés of small-town life. Blumenthal the hometown boy and resident observer went behind the plain fact of gossip to the social processes by which opinion makers created common knowledge—what everyone thought or was supposed to think—and how gossip enforced conformity. Social reality, he discovered, was more complicated than the literary stereotype. Some residents gossiped a lot, others not at all; some were ruled by convention, others quite free. And there were acknowledged ways of evading Mrs. Grundy's iron laws. Blumenthal thus discovered as a resident observer what Malinowski had before him: communal "laws" were not blindly obeyed but in practice were adjusted to the particularities of individual personalities and social situations.[86]

Blumenthal's was an independent discovery of participant observing, in a sociological idiom and to sociological ends. Though Philipsburg and Omarakana were worlds apart, and the roles of native son and resident stranger quite unlike, the circumstances in which Malinowski and Blumenthal discovered an ethnographic and a sociological form of participant observing were much the same. They did so in places that were well removed from the centers and sources of orthodox practices in their disciplines. In both cases discovery was an emergent process, a situated practice growing naturally out of existing practices. Ethnological and sociological surveys afforded the same kind of opportunities to intensify interactions with informants in informal situations, with results that were unexpectedly rich and novel.

HOBO SOCIOLOGIST: NELS ANDERSON

Nels Anderson's hobo investigation is yet another case, and an exceptionally well-documented one, of participant observing growing out of existing practices, in this case social survey and hometown-resident knowhow. Although the project was carried out in Chicago, Hobohemia was a part of the city that for sociologists was effectively a foreign country. Not for Anderson, however: he knew it as an insider, having lived there as a boy with his itinerant family, and from his own experience of hobo work and life. Hobos posed exceptional difficulties for inquisitive sociologists. Single men living beyond the pale of respectably settled family society, they were elusive (most of them passing through from one job to another), wary of strangers asking questions (only cops and social workers did that), and skilled in evasion. One unwary student sociologist lined up some hobos for interviews by giving them food and a night's lodging in a settlement house, only to find them long gone when he returned the next morning to begin work.[87] Hobos were also uniquely unappealing subjects for middle-class sociologists, who like everyone else lumped them together with tramps, bums, and panhandlers—which they were decidedly not. Hobohemia was the one area of Chicago where student sociologists would not go.[88]

"Hobohemia" was actually a misnomer, invented by Robert Park, who liked to think of hobos as romantic bohemians. In fact, they were itinerant laborers, and Hobohemia—a mile or so off Madison Street on the west bank of the Chicago River—was their "Main Stem," where institutions clustered that catered to their needs. These included employment agencies and information exchanges; cheap hotels, lodging houses, and flophouses; cheap eateries and outfitting shops; saloons and bootleggers; houses of prostitution; radical bookstores (many hobos were avid readers); hospitals and clinics; and welfare agencies—everything "to minister to the needs, physical and spiritual, of the homeless man."[89] Hobohemia was a major regional labor exchange—it was estimated that as many as 300,000 itinerant workers per year passed through the Main Stem—and a vast center of cheap lodging for large numbers of temporarily or permanently homeless men (rarely, women and families).[90]

It also had its own complex social order. Anderson distinguished five distinct types and classes of Hobohemians. There was the seasonal migrant worker, who followed a fixed schedule of seasonal work in agriculture and orcharding. Hobos were a second type, who followed no orderly schedule but worked at whatever jobs were available in railroad building and maintenance, in construction work in large developments or in rebuilding after floods or fires; as well as in mining, agriculture, and various trades. Distinct from these two classes were two that did not rely on their own work to live: tramps, who wandered with no fixed aims and without interest in regular work; and bums, who lived by begging or charity and were mainly settled in the city. Finally there was the "home guard." These were permanent residents of Hobohemia, mostly older men who had been migrant workers and who for reasons of age, drink, injury, or broken health could no longer move about, and lived by odd jobs and the intricate art of "getting by." Seasonal migrants were the labor aristocracy of Hobohemia; bums and the home guard, the dependent underclass. The important division was between those who worked, and those who did not—the respectably self-supporting, and those who depended on others. As one knowledgeable hobo advocate put it, "The hobo works and wanders, the tramp dreams and wanders, and the bum drinks and wanders."[91] Welfare agents and the general public tended to lump all together as tramps, or as hobos, or as generic "homeless," in effect erasing hobos' distinctive social identity and economic role.

The intricate, churning, and half-hidden social world of the real Hobohemia was accessible only to someone who knew it from the inside, as Anderson did. Since he derived his understanding and methods of study mostly from his own experience, we need to look first in some detail at his remarkable life.

Nels grew up in an itinerant family. His father, also Nels, was an orphaned Swedish peasant lad when he emigrated first to Germany, where he learned the bricklayer's trade, and then to America, in search of cheap land to farm and a wife to produce many children to work it. He was himself a hobo laborer for a few years before he found a wife and settled, briefly, in Chicago, until news of a great fire in Spokane, Washington, sparked hopes of high-paying work in construction in a region where, he thought, there was good homestead land to be had. Unluckily for the Andersons that was not so, and the growing family (young Nels was born in 1889) spent the next thirteen years shifting from one place to another—ten moves in all—trying and failing to make bad land good. For four months the family was homeless and meandered about the Northwest in a covered wagon. They were taking part, as Anderson later understood, in "the frowsy, seedy last phase of covered-wagon days, hobo families in prairie schooners trying to settle when almost all the good land had already been taken."[92] They could afford only what no one else would want: an unimproved timber claim near Spokane, a sharecropping farm on the Nez Perce Reservation, a rented "ranch" on public school land in Jackson Hole, Wyoming. Mainly they lived in towns, where there was construction work: Lewiston, Idaho, then back to Chicago (1899–1901), before settling for good (in 1903) on a sandy but improvable orchard farm near Traverse Bay in eastern Michigan.[93]

Nels in 1903 was poorly schooled—his stubbornly illiterate father put little stock in education for future farmers—but richly educated by experience of diverse kinds of people and ways of life. He was a curious and observant youth, and from an early age was attracted to marginal people despised by middle-class townsfolk. In the diverse society of Lewiston, he was drawn not to the respectable residents of Main Street but to the rowdy life of "saloon street." There he busied himself collecting empty whiskey bottles for sale back to saloons and befriended prostitutes, who saved bottles for him and gave him tips for running errands. He was drawn as well to the campgrounds on the edge of town where covered-wagon families gathered. Such places were off-limits to Main Street boys; however, Nels's mother firmly believed that her children should learn to know and respect all sorts and take care of themselves, and did not object to the company her son kept.[94]

In Chicago, where the Andersons lived in a tenement near Madison Street, Nels and a friend, Steve, hawked newspapers. Unlike newsboys in the more respectable Loop, they did not just work sidewalks, but hawked papers inside saloons, gambling dens, cheap hotels, dance halls, and houses of prostitution, where Nels again befriended the girls, fetching them groceries and pails of "loose beer" for nickel tips. Within months Nels knew the institutions and people of the Main Stem as intimately as any resident did. His parents again made no moral objection to the company he kept; they were outsiders themselves and welcomed his contributions to the family income. Intelligent and thoughtful, Nels developed a lively interest in whatever work he did and whatever sort of people he found himself among.[95] His early life was in retrospect a worldly, down-scale Parkian education in lived sociology.

Leaving the city and his newsboy business for farmwork and school at Traverse Bay was for Nels an unwelcome loss of variety and independence. His father worked at a nearby iron furnace, earning to buy uncleared land for his four sons' future farms. (It was a father's duty, he believed, to provide a landed future for his children and to make their life decisions for them.) That left Nels, his older brother Bill, and two younger boys to do all the farmwork—after a full day at school and to their father's strict directives. Nels was good in school, making up lost grades by passing two per year. Boarding with a middle-class town family who valued education, he set his mind on one day finishing high school, a plan that found no encouragement at home. His father and brothers disdained education beyond the sixth grade as "just another way of getting out of work."[96] One by one the Anderson boys, and later the girls, left home to find their own ways in various lines of work. None became a farmer.[97]

Bill "skipped" first, in 1906, for hobo work in railroad construction; he was fed up with taking orders from his father and wanted to prove his manly independence. Seeing how easy it was to leave, Nels followed a few months later, collecting his father's paycheck after school as usual but on an impulse keeping it to buy railroad fare to join Bill somewhere on the road. It wasn't the farm drudgery that drove him out, nor the schooling, but rather his social situation in school, where town kids nicknamed him "farmer"—a derogatory term like "hick" or "hayseed"—and where his one good friend was a Native American, George Mamagone, who was another bred-in-the-bone outsider. Nels chose a hobo life to escape a low-caste social identity and win the respect that came from doing a man's work and making his independent way in the world. Three days after leaving home, Nels had found Bill, on a railroad-grading crew near Galesburg, Illinois, and had a job he "had yet to learn and at man's pay"—two and a half times what he would have earned in farm labor.[98] Thus began a varied hobo life that ended in 1921 when a new life began as a professional sociologist.

As with all hobos, Anderson's life revolved around work: finding jobs, learning new skills and work cultures, finding better jobs, moving on.[99] For two years (1906–7) he worked as a mule "skinner" in earth-moving and grading crews, in Illinois and Missouri on the Santa Fe Railroad and in Montana on the Milwaukee Road—the last of the transcontinental lines—with visits home in off-seasons. Hauling dirt around was unskilled labor, but like any work with animals it required experience and know-how; it also got Nels into the more skilled and interesting occupations of blacksmithing and harness making. He especially enjoyed sewing. This period of ready work ended with the panic and depression of 1907, and for four months in 1908 Nels bummed about the West, homeless and looking without success for work in the wheat fields of Kansas and the mines of Colorado and Utah. In this ebb of hobo life, he mastered the practical arts of free riding and evading railroad "bulls" and local cops, and learned the ways of edge-of-town "jungles" where hobos gathered to cook and wash, as well as the know-how of backdoor begging for food and odd jobs—hobos' last resort when down on their luck.

Anderson's life took an unexpected turn from bumming in July 1908, when through ignorance and mischance he found himself ditched from a freight and on foot in the middle of the Nevada desert. He had been on the way to Los Angeles and from there (he hoped) to Panama, to work on the canal, not knowing that the line across the Great Basin was the most inhospitable to hobos in the entire West. Coming upon a tiny green oasis, Nels struck up a conversation with a Mormon farmer at work. Talk led to his lending a hand, and then to an offer of a regular job at a family ranch. There Anderson remained for two years (1908–10), virtually as an adopted member of a Mormon family, who were as devoted to a settled life as he had been to a life on the move. His adoptive family, unlike his own, put a high value on education, and their encouragement and good example with their own children revived Nels's latent intention to finish high school and go on to college. His new family offered to underwrite his schooling, but Nels preferred, as he always did, to work to pay his way.

The financial means for schooling came unexpectedly in 1910, when a flash flood washed out whole stretches of the through railroad line and created a demand for skilled labor so urgent that inexperience was no bar to being hired on the spot, and at a wage that in a year would pay for a year in school. It was also an opportunity to learn the skilled work of a railroad maintenance carpenter. This was not hobo labor but skilled craftwork building trestles, repairing tunnels, doing mechanical and electrical jobs. Whatever needed to be done he learned to do. He acquired his own kit of tools—the means and insignia of craft status—and a new identity as a member of a labor elite. From 1910 to 1917 he alternated between attending high school, then college, and working as a carpenter on the railroad and in mines to pay for it. After an interlude of two years in the US Army in Belgium and France (1917–19), and another year of alternating work and study, Anderson completed his BA degree, at Brigham Young University. He was thirty-one years old. And for the first time in his life he had a definite idea of his future, having been persuaded by a professor, John Swenson, to make sociology his life's work—despite his sporadic and slender academic training. After a summer of crash study, he was again in a freight car—a last hobo shift—on his way back to Chicago and (he hoped) graduate school.[100]

Anderson arrived in Chicago with twenty dollars in his pocket, little knowledge of sociology—he had forgotten most of what little he had been sporadically taught—and an uncertain prospect of admission to graduate study. Most important, he had no job and, for the first time in his life, no sure means of finding one. He had sold his tools before leaving the West and knew nothing of middle-class work or how to compete with middle-class students for what work there was. "I knew how to get in and out of cities," he recalled, "but I had never worked in one."[101] Without the tools, reputation, and connections of a community of work, he was no longer securely self-sufficient. "My tools, like my bed roll," he recalled, "had been links between me and the migratory hobo way of life to which I had become adjusted." He would have to find "other ways of living and earning."[102] But if there was one thing he did know from his hobo experience, it was how to find or create opportunities in unfamiliar and unpromising situations. When jobs were few and job seekers many, for example, he would always look in places where more conventional seekers would not think to go (hence his ill-advised attempt to reach Panama). He also knew that talking with strangers was often the best way to make things happen. A friend who roomed with him just after the two arrived in Chicago remembered Anderson as "a very folksy kind of man" who connected easily with strangers.[103]

Anderson spent his first night in Chicago in a cheap men's hotel on Madison Street, just blocks from where he had lived as a boy. The next day he took a trolley to the alien ground of Hyde Park, where he got hostile stares, to see the university and to scout possible places of employment and neighborhoods where he might afford to live. A redbrick building among all the academic gray stone caught his eye, its designation as "Chicago Home for Incurables" suggesting a place where students would never deign to go. After a night dossing safely in the furnace room of a printing company, he returned to the Home for Incurables, where the head, a Dr. Palmer, gently but firmly told him that he never hired students. He wouldn't hire one either, Anderson unexpectedly replied, thus opening a conversation about his life and army service that resulted in his being offered a job as assistant gardener. That makeshift became a full-time job as Anderson took on needed repairs and improvements—familiar work to a railroad maintenance carpenter—and assisted in the care of incapacitated patients by running errands, writing letters, giving emotional support, and being a needed friend.[104] These were ingrained habits of his hobo life: finding work in odd places, mastering new kinds of work and making himself useful, and engaging fully in the life of whatever community he happened to be in.

His hobo experience also helped get him admitted to the University of Chicago's Department of Sociology. He was well aware that he had neither the knowledge nor the credentials to be admitted, and that fact quickly became apparent in his interview with Albion Small. However, Small was impressed by Anderson's life as free rider and job getter, and especially impressed, and amused, by his landing a job in the Home for Incurables. He had sent his friend Dr. Palmer many students, but every one of them had been sent back. Small allowed that he would be pleased if Anderson would take a class with him in his first term. Anderson never forgot the gracious "violation of procedure."[105] By conventional academic rules it was doubtless a violation, but not, I think, in the department of Robert Park, where worldly experience and interest were no less qualifying than academic credentials.

IN HOBOHEMIA

So Anderson found himself once again working and studying, not in alternation now but at the same time; and that was a problem. The expanding work with his "incurables" left little time for study and even less for the informal student life through which he would learn the customs and idiom of his new working community. Finding time for the fieldwork of term papers was a particular problem. For some courses he was allowed to substitute his hobo knowledge. He did a small case study of a man whom he knew only as "Mugsy," a respected citizen of the Main Stem who pursued his occupation of specialist "highdiver" pickpocket outside the city limits. In a summer project in 1921 he interviewed some four hundred hobos, presumably by questionnaire. However, he struggled to learn enough basic sociology to formulate a project for his MA thesis.[106] What he needed was work that was both sociology and a living.

The hobo project came to him serendipitously through his hobo experience, when a social worker friend alerted him to a lecture on Chicago's homeless men by the eccentric social activist, ex-tramp, and physician to the poor, Dr. Ben Reitman. Reitman belabored his audience of social workers for treating the homeless as statistics and not as individual human beings. It was a rambling discourse, more entertaining and provocative than informative. In the vigorous dispute that followed the talk, Anderson reproached Reitman for ignoring the working lives of hobos outside of Hobohemia and for not adequately distinguishing hobo workers from nonworking tramps and bums. Discussion became a dialogue as the audience slipped away, and continued at Reitman's invitation over pie and coffee, when he revealed a plan for a full-scale study of homeless men and asked Anderson if he would design and run it. Reitman promised (and delivered) funding by Chicago's Department of Public Health and oversight by United Charities. And Burgess and Park's approval of the project for Anderson's MA thesis gave academic standing to what was for its sponsors a remedial social survey and for Anderson a paying job.[107]

The homeless survey, once Anderson had eased himself out of the Home for Incurables, resolved the problem of fieldwork. He was, as he put it, "'in the field' and . . . in the midst of material which was both real and pertinent."[108] For the first time in his life he had paying work that advanced him toward a definite life goal, though knowing his work would be judged by sociologists was "a frightening prospect."[109] Nor were his teachers much help in a practical way. Robert Park's only advice was to "write down only what you see, hear, and know, like a good newspaper reporter." However, from a life of learning on the job Anderson was already equipped with "a capacity for interviewing and a capacity for reporting what I had seen and heard."[110] Although the purpose of Reitman's project was remedial, its means were to be social-scientific. The idea was not just to gather the facts of itinerancy and its inevitable downward course, as in the usual social survey, but to discover its situational and psychological causes: "the human nature of the migratory casual worker, and . . . the economic and social forces which have shaped his personality," as Burgess put it.[111] Burgess could not tell Anderson how to accomplish that; but Anderson knew to be guided by his personal experience of hobo society, and Burgess approved.[112]

Anderson pursued several modes of inquiry in the hobo project. One was a sociological survey of Hobohemia as an urban natural area that inventoried and mapped the Madison Street institutions that served (and exploited) hobo needs, with a detailed account of the various grades of lodging houses, plus descriptions of their denizens' many shifts of "getting by." This survey anticipated the studies of natural areas and occupations in Chicago's Near North that would soon become the trademark of the Chicago school.[113] A second line of inquiry was a traditional social survey of the so-called hobo problem, meaning aspects of hobo life that for social workers and civic agencies were problems to be treated—poverty and dependency; health, sex, and venereal disease; begging and petty crime—with recommended remedial measures.[114] For this line of study Anderson drew heavily on records in agency and civic archives and the large literature on homelessness, as well as on his own street work looking into lodging houses, taverns, and places of entertainment. In a brief venture in participant observing, he checked into a fleabag flophouse (Hogan's Flop) to experience for himself the lowest lodging, but left in a hurry in the middle of the night when he was beset by legions of aggressive bugs.[115]

The third and most original line of Anderson's project was his study of the varieties of hobo workers and the psychosocial dynamics of hobo lives. As Anderson put it, "Why are there tramps and hobos?" Why do men leave home to become itinerant workers, and why do some remain unsettled despite the predictable downhill slide with age into dependency and indigence? Are the causes personal and within, or in the circumstances of their lives? Young hobos tended to blame circumstance, whereas old ones blamed some failing in themselves.[116] It was in this part of the project that Anderson relied least on the new and unfamiliar techniques of sociology or social work and most on his own personal experience, in which he felt secure. The sociological survey of Madison Street institutions was an "afterthought," he recalled.[117] Mainly he used his personal knowledge of hobo life to engage directly with individual hobos in their own social habitats.

The principal tool of this line of inquiry was the conversational interview—a quasi-ethnographic activity that Anderson did not have to learn or invent, because for him it was second nature. Hobos had a distinctive manner of speaking and conversing and were accomplished storytellers. On the road and in jungles their stories were generally not about personal and inner lives, but about the practicalities of life on the road: stories of cleverly evading police and railroad "bulls," and of being down-and-out and begging for food; stories that displayed competence and worldly know-how, and of fortune, good and bad. The brief and ever-shifting associations of hobo work gave them no reason to take an interest in each other as persons. On the road and in jungles few knew their fellows by name, and they gathered with no hellos and moved on with no farewells. Their talk was of shared concerns of job seekers: news of good and bad places of employment, current conditions on the road, places to seek out or avoid. The forms of hobo oral culture were thus designed to exchange information vital to life while discouraging personal intimacy. As Anderson put it, hobos "lived closed lives and granted others the same privilege."[118] However, an observer who knew the codes of interaction could easily and inconspicuously turn these conventions to sociological ends. In his year of fieldwork in Hobohemia, Anderson was by this means able to gather life histories of over sixty hobos.[119]

To stay close to his subjects and keep his cover (and to save money), Anderson took a room in a working-class hotel just half a block from the tenement where his family had once lived—an area he knew intimately from his newsboy days. To engage his neighbors in conversation, he had only to hang out and talk to them in their own idiom and manner. He knew to avoid direct questions about personal histories—these were guaranteed conversation stoppers. Instead, he began with casual chat about work and life on the road, then gradually steered the conversation to more personal matters: reasons for leaving home and then for staying out, hopes and disappointments, personal failings and misfortunes. In Anderson's own words,

> Wisely or not, I began with informal interviews, sitting with a man on the curb, sitting in the lobby of a hotel or flop house, going with someone for a cup of coffee with doughnuts or rolls. . . . I was at home in that area . . . and . . . equally at home among the inhabitants of the area. . . . I could talk without uneasiness about having come from one place or other in the West, of having done one kind of work or another. It was an advantage to be able to talk about the types of work men in that sector of society do, and work talk turned out to be a productive inducer of general conversation. Even men who at the time of meeting were living entirely by begging had their work histories of one type or other, especially if the beggar was an older man, often he was wont to relive his work life gladly. An occasional question, casually asked, would prompt him to expatiate about home, family, how he began moving about. An observation from me that I had wasted time bumming around when I could have gotten ahead had I not left school usually brought like responses. I discovered that one sitting next to someone else can effectively start conversation by thinking aloud. It invites attention and one needs but to come out of his reverie, tell of the thoughts in mind.[120]

It was, to say the least, "a type of interviewing which was outside the range of experience, knowledge and understanding" of his fellow students.[121] For Anderson it was a continuation of hobo life to different ends. And it produced a trove of inside stories of the patterns and variations of hobo lives, and empirical explanations for why there were hobos.

The reasons why men (usually young men) left home were as varied as their personalities and social situations. For some the reasons were economic, when seasonal or periodic unemployment and local failures of farms or factories drove them to seek temporary employment far afield. For some it was human relationships gone sour: with wives or girlfriends, bosses, or bossy fathers. In farm families, children were typically expected to work and obey even into early maturity, despite attractive opportunities for independent work in urban industries. Other common causes were embarrassment and shame at some failure at home or work, peer disrespect and ridicule for a personal oddity or blunder, and ethnic or racial prejudice. Or plain restlessness and desire for adventure and a chance to display manly independence, which were common desires in young males, fostered by new opportunities for long-distance rail travel and a regional culture of mobility. Social workers and psychiatrists tended to see restlessness as a kind of individual psychopathology. But in Anderson's experience, though many hobos were square pegs or impaired in some way, many were quite normal; and in IQ tests hobos tended to be somewhat above average.[122] Anderson's account is full of little stories from his records of interviews of hobos talking about their upbringing and lives on the road, and these life stories are the empirical substance and analytic frame of Anderson's sociology.

In all the variety of what pushed individuals to leave home and stay away, there was one common element: the pull of hobo work. It was not a last desperate resort for those disappointed by home and village life, but an attractive alternative to the often monotonous and restrictive work of farm and factory. Hobo work was varied and generally easy to get, paid better than farm labor, and gave young men opportunities to display manly independence—a key virtue of hobo culture. Hobo working groups were generally egalitarian and accepting of personal oddities, in contrast to the conformity of small-town culture at the time. If bored or dissatisfied with the work they were doing, hobos could quit and usually be confident of finding other work quickly, thanks to the well-developed system of main-stem employment agencies and informal hobo grapevines. Job mobility was for hobos a valued sign of competence and self-support. The variety and the accessibility of hobo employment were major factors in turning first-time hobos into lifelong ones. As Anderson put it, the pull of hobo work selected out the most restless men and made hobos of them.[123] It drew them to the railroads, mines, and harvest fields of the Great West and kept them from returning to a settled way of life. For those who were experienced and knew the system, taking another hobo job became the path of least resistance, and one thing led easily to yet another.

Anderson was not the only writer on homelessness to see that work was central to hobo life and culture. But he was the one who made it the central theme of his sociology and gave it the richest support of in situ firsthand observation. He did not dispute the common idea that itinerant workers were victims of the exploitive labor markets of modern industrial capitalism. But he also gave hobos the dignity of being a special kind of pioneer: "They have contributed more to the open, frank, and adventurous spirit of the Old West than we are always willing to admit," he wrote. "They are, as it were, belated frontiersmen." It was a canny insight into a distinctive and passing labor system.[124]

The reasons why Anderson made work so central to his sociology of hobodom lie in a particular feature of his own life experience. From an early age Anderson was fascinated by work and workers, and he developed an inveterate habit of entering into working cultures. The chapters of his life history are marked out by particular lines of work: farming, newspaper hawking, mule skinning and grading, maintenance carpentry, blacksmithing and sewing, nursing, social science. This fascination derived not just from the satisfaction of mastering new skills but also from the emotional and intellectual rewards of entering into and understanding the varied ways of working life: learning the tacit rules and values of worker communities, their customs of social demeanor and relations, their moral principles of mutual rights and obligations—what Malinowski called the "charters of native institutions." There may have been deeper roots as well, in the Anderson family's repeated experience of being socially marginalized. Anderson's serial attachment to communities of work was perhaps a conscious or unconscious effort to experience the acceptance and respect that he and his family never quite had. Anderson hinted as much in a recollection of his mule-skinning days: "That work and those men caught my interest. It may well have been that I felt myself accepted and I was reciprocating by accepting them and their way of life."[125] That insight plausibly extends to his acceptance of and by saloon-street denizens, hustling newsboys, emergency carpenters, Mormon farmers, middle-class collegians, and—last and most lastingly—professional sociologists.

Anderson was always conscious of the social side of work, even as a boy of ten selling newspapers with Steve: "much of the learning was the multitude of gestures, signs, quick answers, and, of course, the tricks of the trade [that is, petty cheating]," he recalled. "I was entering an occupation that paid off later. Indeed, I did not realize until years later how important it was to be."[126] In his first work as a hobo, he was fascinated by the personalities and manners of experienced hobos. Along with the techniques of steering a mule team, he also mastered mule skinners' social codes: their dress, habits of social interaction, specialized jargon, and shifts of itinerant life. "Like the student who learns the 'theory' of his occupation in school," he recalled, "I learned how the hobo behaved, or should not behave, in town, how he went about from place to place on freight trains, how he evaded train crews and railroad police, and how he found his work."[127]

Anderson especially admired older "all-round" hobos. He liked their pride in self-sufficiency and their reluctance to stay too long on one job, lest they lose the omnicompetence and mobility that gave them self-respect and distinguished them from idle tramps and bums. After just six months on his first job, Anderson, too, began to feel "self-conscious" about holding on so long and worried that he was becoming "what the hoboes call a 'home guard.'"[128] Moving about quickly and safely by free riding, and meeting the challenges of new kinds of work and working groups, he felt "a greater identity with my ideal of the hobo." He learned hobo work not just to make a living but for the satisfaction of working as expertly as the most experienced and able hobos did.[129] He liked knowing and performing the codes and behaviors that gave human communities their distinctive characters.

It was the same with academic work and culture. As a beginning student at Brigham Young University, Anderson took pains to observe collegians' way of life and make it his own. A stranger once again at the University of Chicago and a "poor mixer," he was unable to take part in students' conversations, which were mainly about their own projects. When his own work on hobos came up, it was usually "in a joking context." But he listened and observed, knowing that "by and by the psychological barrier would come down" as he learned the appropriate use of sociological words and concepts.[130] "I felt that a craftsman was a better craftsman if he knew and obeyed the roles of his trade," he recalled, "and I expected the same would hold for students and professors."[131] Anderson's new life in sociology was in many ways a continuation of his old one, just as Robert Park's sociology was a continuation of his life as a journalist and activist. For both, worldly life provided the stuff and the observing habits of social science.

This connection of life and sociology was not hindsight imagining. Anderson's gift for thinking sociologically caught the eye of his sociology professor at Brigham Young, John Swenson, who urged him to give up his notion of becoming a lawyer and to become a sociologist: "My background, he said, 'would be useful in law, but valuable in sociology.'" After giving it some thought, Anderson agreed.[132] This is not to say that as an observant, thoughtful man of the world he was a kind of everyman sociologist. He was not. He was a working man. But he entered into communities of work in much the same spirit that ethnographers and sociologists entered into communities, to understand their operating principles—just to different ends. Becoming a sociologist was the last of his hobo shifts. As Anderson the hobo was an observant participant, so was Anderson the sociologist a participant observer. Habits developed in the shifts of hobo life became, in Hyde Park and Hobohemia, an inside science of engaged observing.

Anderson did not use his own life story in *Hobo* (except for the tale of Hogan's Flop), so his readers at the time would not have known how intimately his life and science connected. But if you've read his later autobiography, his presence in the book is clear enough. In the stories of hobos' lives that Anderson recorded in Hobohemia—some quite similar to his own, though displaying downward paths that he escaped—we see the total situation of the time and place in which Anderson's own story was one variant. His own life experience is present in his sociology not just in the techniques of access and interview but in its foundational perceptions of the meaning of hobo life. *Hobo* was in a way a group autobiography. And the originality of his science lies in his idiosyncratic life.

CONCLUDING THOUGHTS

Anderson's hobo project—one of the first of the Park-Burgess period—was in many ways a singularity: in its subject and circumstances, and in the life experience that went into it. Not surprisingly it has been hard to pigeonhole. Anderson's classmate Pauline Young cited it in her 1939 handbook of social survey as an example of "uncontrolled participant observation," which removed "social and mental distances through intimate participation"—that is, ethnography.[133] But upon learning of Young's statement (in 1965), Anderson gently declined the compliment, pointing out that he had not been a stranger in a strange land, as anthropologists typically were, but at home among his hobo subjects and like them just doing a job: "I did not descend into the pit, assume a role there, and later ascend to brush off the dust. I was in the process of moving out of the hobo world. To use a hobo expression, preparing the book was a way of 'getting by,' earning a living while the exit was under way. The role was familiar before the research began. In the realm of sociology and university life I was moving into a new role."[134] Jennifer Platt (in 1983) took Anderson's statement as evidence that his fieldwork was not participant observation, but a stock Chicago case study.[135] Yet if Anderson was not a participant observer in his social role, the method was "faithfully followed in my work," as he later wrote.[136] What he did was, so to speak, participant observing *avant la lettre*.

Or we could place his project in the tradition of practical social survey. Ben Reitman saw its purpose as remedial, and that is how Anderson took it and how (mainly) he wrote it up. He was taken aback when Park announced that his report would be published as the first of the department's series of monographs in sociology. He had not written it as sociology, he protested, but as a guide to programs in aid of the city's homeless.[137] Yet *Hobo* is no less sociology, as his teachers recognized. Or again, we could with hindsight see Anderson's project as an unwitting work of hometown sociology. Anderson was, after all, a hometown boy in Hobohemia, although he was almost certainly unaware of that indigenous sociological practice. The polymorphous quality of Anderson's hobo work is indicative of the inchoate character of sociology in Chicago (and elsewhere) in the early 1920s, when its field practices were still evolving out of older indigenous practices and new grafts from journalism and other occupations of everyday life. That was so especially in Chicago, where Park's capacious vision of an urban sociology nurtured a branching bush of field practices, and individual projects took particular shapes in the doing of them, not from any settled academic forms.

This last point is highlighted by the story of Anderson's MA examination, in which he failed, once again, to answer most of the sociological questions put to him. Yet when he was called back to the examination room to hear the examiners' verdict, "Professor Albion W. Small pointed to the street [and said,] 'You know your sociology out there better than we do, but you don't know it in here. We have decided to take a chance and approve you for your Master's degree.' "[138] In Robert Park's department it was the sociology "out there" that mattered, and Anderson had been "out there" observing and entering into the lives of others long before he was an apprentice sociologist. No disciplinary label exactly fits his sociology, because a vital source of it was his own life.

Although *The Hobo* enjoyed considerable celebrity (it is still in print), neither it nor Anderson's bootstrap personal narrative became models of a resident sociology, as Malinowski and his Trobriand monographs did in social anthropology. At a time when formal schooling was becoming the exclusive path to careers in social science, Anderson's life history was just too heterodox to be widely emulated. Nor were hobos a model subject in an academic culture in which scholars' subjects tended to be read as endorsements of ways of life, and any suspicion of a taste for low life was likely to be fatal, as it had been for W. I. Thomas. Anderson's subject and life history effectively barred him from a career in academia. He tried to shed the unsought identity of "hobo sociologist." He declined many of the invitations he received to lecture on his book, and took as the subject of his PhD dissertation, at New York University, the historical part of a historical sociology of New York City's slummy urban edge as it moved from lower Manhattan up to its then-current position in east Harlem. It was the safest part of a safe subject, and the part that could be pursued in libraries and archives, not in slums.[139] It was as if Anderson wanted to erase his unorthodoxy by playing strictly by the rules of upscale academic sociology. Yet full acceptance in that community of work was denied him—a first for him. He spent his professional life as a government official in the United States and Europe, serving the uprooted, homeless, and displaced, whose numbers and needs were extraordinary in the Great Depression and in World War II's aftermath of mass expulsions and ethnic cleansing. Only after his retirement from public service was he offered a professorship, by the University of New Brunswick.[140]

In academic sociology, meanwhile, older makeshifts of social survey and hometown sociology were receding into history, as new generations of sociologists and ethnographers looked over the disciplinary fence to see what the other was doing.

CHAPTER 4

CORNER SOCIOLOGY

WILLIAM WHYTE

If Nels Anderson's *Hobo* and Albert Blumenthal's *Small-Town Stuff* were (in hindsight) harbingers of resident observing in sociology, William F. Whyte's *Street Corner Society* marks its open arrival. Published in 1943 and augmented in 1955 with an engagingly personal account of Whyte's field methods and experience, the book was a long-term best seller and an influential model for subsequent studies of small groups by sociologists and practitioners of the new sciences of industrial and human relations.[1] Jennifer Platt is again a dissenting voice: what Whyte did was in her view too mixed with sociometric and other techniques to be true participant observing. It was only Whyte's autobiographical afterword, she suggests, that gave *Street Corner Society* the hindsight appearance of a "virgin methodological birth."[2] Impure Whyte's practice may have been, yet it was unmistakably participant observing. He resided intimately among his North End neighbors and took part as one of the boys in the activities of the clique of young men who became his main subjects. He even experienced with his own body the psychosocial mechanisms that shape group structure and behaviors—a depth of affective participation rare even among anthropologists.

Whyte's book was to be sure no "virgin birth." Like any new practice it was fashioned out of familiar current practices. Much had changed in American sociology in the twenty years between *Hobo* and *Street Corner Society*. The really big news was the advent of the radically desituating practice of random-sample polling, the so-called *new* social survey (no relation except in name to the old "social survey"), which became a mainstream practice in the mid- to late 1930s and near hegemonic in the postwar decades. No less significant, however, was the simultaneous increase in traffic between sociology and social anthropology. What had been a well-defined boundary became in the 1930s a porous border zone of mixing. Anthropologists, on one side, took to investigating contemporary Western societies, hitherto the exclusive domain of sociologists; while on the other side, sociologists adopted ethnographers' methods of participant observing as their own. One result of this border-zone exchange was a distinctive practice of "community study," in which both anthropologists and sociologists took part. The subjects of community study were whole, and sizable, communities—the familiar subjects of both sciences. However, the mixed methods of community study could be applied to other social groupings as well: like small groups. And the striking early achievements of community study—especially Robert and Helen Lynd's famous *Middletown*—invited others to apply resident observing to modern societies in their own particular ways. This was the context of practice in which Whyte's study of small groups in "Cornerville" took shape.

Whyte did not set out to create an ethnographic sociology of small groups. Neither sociologist nor anthropologist, he went to Boston's North End with the idea of doing a full-scale community survey. He knew about participant observing from the anthropologist Conrad Arensberg, a friend and colleague and author of a model community study. Yet only in the middle of his project did he unexpectedly realize that he was doing a kind of ethnographic sociology, not of a whole community, but of the small groups with which, for convenience, he had begun. Like Malinowski's discovery of participant observing, Whyte's was an unexpected outgrowth of an orthodox mode of practice. As Malinowski's grew out of traditional ethnological survey, and as Nels Anderson's and Albert Blumenthal's resident observing grew out of sociologists' practices of social survey and hometown study, so did Whyte's small-group sociology take shape within the mixed practice of 1930s community study.

SOCIOLOGY AND ANTHROPOLOGY: A MEETING OF WAYS

There is as yet no systematic history of relations between the sibling sciences of sociology and anthropology between the wars. But in the United States the pattern of an increasing convergence of subjects and method is clear enough. As anthropologists ventured into sociologists' domain of contemporary societies, sociologists took up ethnographers' techniques of residing and participant observing. A precondition of this conjuncture was ethnologists' abandonment of their historicist belief that "primitive" societies were relics of early phases of human social evolution. So long as their aim was to reconstruct that deep history, modern societies were unusable subjects, like traditional societies contaminated by European contact, only more so. For the same reason anthropologists' monographs on premodern societies were unusable for sociological studies of how contemporary societies operate.

This mutual exclusion changed when ethnologists abandoned historicism for a functionalist interest in the operating principles of all kinds of human society. Modern and modernizing societies, once taboo, became available and attractive subjects for ethnographers, and practical considerations of training and career heightened their appeal. Their variety and accessibility made them the subjects of choice as global depression and world war made travel to distant and exotic societies difficult or impossible, and when universities in the postwar years were flooded with aspiring anthropologists chasing a dwindling supply of unsullied and unstudied premodern societies.[3] Modern societies thus became an open frontier for ethnographic investigation. Anthropologists' intrusions into sociological subjects also made their practice of participant observing more accessible and inviting to sociologists, who were already moving, in their own ways, in that direction.[4] A border zone of cultural mixing thus formed, in which sociologists and anthropologists could discover shared interests in questions of social structure and cultural change, and a shared commitment to methods of resident observing.[5] Institutional impediments to cross-border sharing did persist, in the vested interests of academic departments in the subjects and methods that gave each their right to students, space, posts, and resources. For this reason, borrowing was at first easier on the personal level than on the institutional level, as we will see.

A particular bridge between the two sciences was the phenomenon of culture contact, or acculturation: the process by which premodern societies were transformed or obliterated by contact with Europeans and European ways.[6] Anthropologists could examine the process of acculturation in premodern societies using methods that were familiar and securely anthropological; and from there it was but a short step to an ethnography of modern societies. In 1929 Malinowski called upon ethnographers to create a field of the "changing native"; and in the 1930s, impelled by the practical needs of indirect colonial rule, British anthropologists moved decisively into studies of culture contact.[7] In the United States, efforts to give acculturation official recognition as a subject for anthropologists became official in 1935, when the Social Sciences Research Council constituted Robert Redfield, Ralph Linton, and Melville Herskovits as a committee on acculturation. The editor of the journal that published their report nervously asked the American Anthropological Association to rule on the propriety of giving space to sociologists' concerns in a core disciplinary journal, but it proved to be uncontroversial. Gatekeepers neither opposed nor endorsed the gambit.[8] Two years later Herskovits systematically laid out a range of topics, approaches, and modes of analysis for ethnographers—an updated *Notes and Queries* guide to ethnographers of modern societies.[9] There was lingering uneasiness about erosion of tribal academic identity. A young anthropologist investigating the coming of modernity to a rural Missouri town recalled that many anthropologists at the time (1939) regarded it as odd and even "unanthropological" to study one's own culture.[10] But no one stopped him. Adapting ethnography to contemporary subjects was itself a kind of acculturation in academic culture—discomfiting yet desired.

Although Herskovits mainly addressed anthropologists and considered only non-Western societies, he hoped that sociologists would join in the study of acculturation, and it was not an unrealistic hope. Immigration, assimilation, and class and ethnic mixing—core subjects of sociology—were instances of acculturation, though sociologists did not use the ethnographic term. A context was thus prepared for rapprochement. An early sign of mutual interest was the publication in the 1920s of a spate of books by American anthropologists, including Clark Wissler and Alfred Kroeber, that were intended to make ethnographic concepts accessible to non-anthropologists. These books produced "a sort of ferment in sociology," the sociologist Jesse Bernard reported, though it was the theoretical concepts of "culture" and "culture areas" that were most likely the leaven in that ferment, not ethnographic field methods: of these Bernard made no mention.[11] When Robert Park asserted (in 1925) that the "patient methods of observation" that anthropologists applied to premodern societies would be even more fruitfully applied to modern urban ones, he cited anthropologists (Franz Boas, Robert Lowie) whose fieldwork with native Americans was more salvage and inventory than participant observing.[12] Sociologists thus entertained the idea that borrowing was a good thing to do before they knew exactly why or how to do it. Was ethnographic practice really practicable or worth the effort for them? Arthur Wood, who studied communities and advocated hometown sociology, warned uneasily in 1928 that for "participant observers" (his scare quotes) to really know a community would require the years of residence described in ethnographic monographs. And no study by a sociologist, he thought, could ever be half so revealing of community life as the novels of Hugh Walpole—a rather deflating admission.[13]

Yet sociologists and anthropologists alike were by then looking and moving across the disciplinary divide. An indicator of this growing engagement was the appearance in the 1930s and 1940s of a new kind of community study, in which sociologists and anthropologists (mostly the latter) applied ethnographic methods to sociological subjects. By my unsystematic tally at least two dozen such studies were carried out between the mid-1930s and about 1950. These varied in subject and style, but a common concern was how traditional societies were adapting to economic and cultural modernity—that is, acculturation. The mix of disciplines varied. Some studies by anthropologists of peasant villages in Central and

South America and East Asia were close to traditional ethnographies. But subjects closer to home included rural Ireland, mixed-race communities in Chicago and the American South and Southwest, mixed French and Anglo villages in Quebec, a Navajo pueblo, a Missouri country village, and industrial towns in New England and the American Midwest.[14] Robert and Helen Lynd's survey of Muncie, Indiana, aka Middletown (1929), was an early and atypically large-scale example of the type, as William Whyte's of Boston's North End was a late and atypically small-scale one. Just three of these community studies were carried out by sociologists—Everett Hughes, John Dollard, and August Hollingshead—all of them either trained or employed at the University of Chicago.[15]

It is unsurprising that anthropologists outnumbered sociologists in these hybrid ventures. For anthropologists contemporary communities were vanguard subjects and a smart career choice, whereas in sociology in situ investigations were being displaced in the vanguard by the "new social survey" of random-sample polling and statistical analysis. From a pollster's point of view, everything that could be got in face-to-face interviews could be produced more cleanly and efficiently, and on a much larger scale, by telephone or mail with standard questionnaires. Intensive investigation of particular cases was easily dismissed as an indefensible waste of time and money and an unwise career choice. "It became known, with a certain condescension, as the 'anthropological' method," Everett Hughes recalled, "usually done with a certain impatience, since it delays the real work of 'administering' the questionnaire."[16] So as polling survey became the modern, fast-track road to successful careers in sociology in the late 1930s and 1940s, resident-observing sociology became a less certain byway.[17] Whereas a disciplinary commitment was an incentive to ethnographers to take up community study, with sociologists it was the other way around. There is some significance, then, in the fact that the most influential community studies, Robert and Helen Lynd's of "Middletown" and William Whyte's of "Cornerville," were carried out by individuals with no institutional commitments to either science.

The Lynds made much of anthropological concepts and field methods in their Muncie project, and were well read in the ethnographic literature—Wissler, Rivers, Malinowski, Radcliffe-Brown were all cited and used. The Lynds adopted Rivers's six basic social functions, as well as anthropologists' culture concept.[18] However, they deployed anthropological concepts to distinctly sociological ends. The Muncie project originated as a social survey of religious attitudes with the practical aim of strengthening religious life by understanding and strengthening the psychosocial forces that inspired religious commitment. However, Robert Lynd diverted the project to a sociological and critical investigation of how a traditional country town was swept into the modern world of national corporations and markets, and mass consumer culture. Neither sociologist nor social worker, Lynd was a progressive social critic and activist with a degree in divinity. Unfettered by any claims of disciplinary identity, he could use to his own sociological end whatever science seemed most useful; and that science for him was anthropology.[19]

The Lynds' field methods were up to a point residential. They and their secretary, Frances Flournoy, lived in Muncie, in private households or apartments, for the full eighteen months of fieldwork (January 1924 to June 1925). They took part in local life, broke bread with people of all sorts, and attended public events—church services, school assemblies, court sessions, political rallies, labor meetings—taking surreptitious notes in a "constant interplay of spontaneous participation and detached observation." Four-hour interviews of selected individuals "elicited what is believed to have been almost complete frankness of response." From this observational material, hypotheses were drawn and then tested in interviews of a spectrum of businesses and working-class families, using standard questionnaires.[20] The project was thus a true hybrid. In a review of *Middletown*, Ernest Burgess called it a product of "participant observers" and "anthropological method." But as Jennifer Platt noted, it made as much sense to see it as a late work in the tradition of social survey.[21] In fact, it was both: a sociological survey framed by anthropological concepts and employing, along with survey techniques, ethnographic methods of firsthand observing to give human meaning to generic survey facts. Robert Lynd became (for a time) a professional sociologist, at Columbia University; and *Middletown* became for sociologists a model of community study.[22] Thus did ethnographic concepts and methods enter indirectly into the domain of sociology.

Direct cross-fertilizing of sociology and anthropology between the wars took place mainly at the University of Chicago. The two disciplines there were conjoined uneasily in one department, an early makeshift that lasted until anthropology at long length won its independence in 1929. The anthropologist and archaeologist Fay-Cooper Cole, appointed in 1923 to turn a museum program into a modern social science, had an active interest in immigration, racial mixing, and culture contact in Latin America. Yet the department remained quite traditional, and the two degree programs kept a wary distance, despite—or perhaps because of—Robert Park's somewhat imperious belief that functionalist anthropology was "nothing more or less than sociology, with the qualification that it is mainly concerned with primitive peoples."[23] Institutional issues of academic prerogative and turf likely impeded cross-border traffic.

Chicago sociologists and anthropologists connected more easily on the personal level, where individuals of like mind could quietly cooperate without stirring up departmental protectionism. Thus Park eased a young Robert Redfield out of a brief and unhappy career in municipal law and into a happy one in anthropology, then helped him shed his outmoded Boasian historicism for a synchronic sociological anthropology, and provided vital support for his unorthodox work as a graduate student and (from 1927) a faculty colleague. Redfield married Park's daughter, Greta, shared an office with sociologists Herbert Blumer and Louis Wirth, and formed close family and professional ties with sociologists Everett Hughes and Helen McGill Hughes. The Redfields and Hugheses together with their students constituted a border zone of reciprocal interest and borrowing.[24]

Redfield's pioneering ethnography of Mexican village life is a good example of how this mixing worked. The project began as a student field exercise in Chicago's small Mexican community, where Redfield mapped institutions, tabulated occupations, and explored life histories and attitudes in intimate interviews—a standard Chicago sociological survey. The exercise lasted just eight months, however. Redfield's subjects were mobile agricultural laborers and in town only between jobs, which made them difficult subjects for sustained ethnography. Nor did Redfield ever quite succeed in winning the trust of an ethnic group convinced (no doubt rightly) that white Americans regarded them as inferior people. He concluded that to understand his subjects' lives and views he would have to either join them in their seasonal circuits of work, which was impracticable, or observe them in their own home villages in Mexico, where he would be the stranger who didn't fit. So in 1926–27 Redfield spent nine months as an ethnographer in Tepoztlán, at first with his family and then, when revolutionary activity threatened, by himself. Though Redfield was there as an anthropologist, his anthropology was not the customary inventory of cultural traits but a functional sociological study of the process and experience of acculturation to a modern economic way of life.[25]

What worked in the field, however, did not necessarily work institutionally. Back in Chicago Redfield found that his project required delicate negotiating of departmental requirements. His dissertation adviser, Fay-Cooper Cole, wanted him to produce a purely descriptive ethnology that could be used in dealing with Chicago's transient Mexican laborers. Robert Park, on the other hand, wanted a sociological analysis of how a peasant community struggled to adapt to an invasive capitalist economy. So Redfield did a bit of both: a salvage ethnology of a traditional peasant way of life, and a sociology of acculturation. His book left neither side fully satisfied. However, it proved to be an influential model for community studies that would, like his own, use both sciences to understand the troubled process of acculturation to modernity.[26] Redfield's subsequent ethnographic survey of mestizo villages in Yucatán likewise accommodated the divergent requirements of science and departmental interest.[27]

To the north, French Canada was another traditional society in transition that attracted Chicagoans concerned in one way or another with cultural change. Horace Miner, a recent graduate in anthropology, selected for his study a rural French village as little altered by modernity as he could find, and produced a descriptive salvage ethnology of a relict culture.[28] Everett Hughes, in contrast, chose a small Quebec town, "Cantonville" (Drummondville in real life), that was historically French and rural but, with the recent construction of two textile mills and an influx of an Anglophone social and business elite, was in the throes of social transformation. (Redfield's friendly help was acknowledged.) Hughes intended Cantonville to be the first of a series of studies of acculturation in contemporary communities.[29] In the event, however, his first such study was also his last.

Upon returning to Chicago, in 1937, Hughes devoted himself to reviving the Park tradition of student fieldwork, which in the 1930s had suffered a decline in activity and academic appeal. Students, he recalled, loved big abstract ideas and disdained the messy particulars of real life as "banal, trivial, and often misguided." To get them to appreciate the value of sociology in situ, he urged them—in words oddly echoing Albion Small's appeal for hometown sociology—to take their own life histories as "opportunities for research that might produce the enlightenment students sought without alienating them from their past." Hughes was proud of his own Ohio farmer forebears and saw no shame in any way of life.[30] Like Park, he strongly identified with marginal social groups, and he urged his students to do likewise—as he had done as a young man, working in ethnic working-class communities of north-woods Wisconsin and industrial Chicago. As he put it, a good sociologist "has to be close to people living their lives and must be himself living his life and must also report."[31] Hughes thus envisioned an ethnographic sociology not of entire towns, like Drummondville or Muncie, but of small groups. That union of life and science became the guiding vision of the so-called second Chicago school—Anselm Strauss, Howard Becker, Erving Goffman, and others—who were in their different ways participant or at least situated observers of occupational or cultural groups.[32]

Howard Becker's path to sociology exhibits the continuity between life and science that Hughes sought to cultivate. Becker was a precocious undergraduate at the University of Chicago in the late war years, more or less as a cover for what he loved best—playing jazz piano semiprofessionally in local taverns and strip joints—and hoped to make his life's work.

He was thinking to major in English (because he liked to read) when he discovered St. Clair Drake and Horace Cayton's great community study of African-American South Chicago, *Black Metropolis*, and realized that he, too, could be an anthropologist—at home and without having to suffer the deprivations of life in some remote and alien outback. All he had to do was to make what he did as a jazz pianist double as anthropology: "Wow! If I just wrote down what I was doing at night, just what everyone said and what I observed, then those were field notes." So he signed on for MA work in sociology and discovered (in a field practicum) that he had a taste and a knack for ethnography. Nudged by Everett Hughes, he drifted into a PhD program and a career as a resident sociologist of outsider groups. Pot smokers, artists, and workers in low-caste occupations were some that caught his interest.[33] It was a science *of* and *in* everyday life.

Although it was overshadowed by the postwar juggernaut of large-scale, big-data team research, the second Chicago school was arguably the most vigorous and fruitful outgrowth of interwar community study. The revival of participant-observation research that Herbert Gans remarked on in the 1960s was likely, as Gans surmised, the demographic bulge of students of Everett Hughes beginning to send their own students into the field.[34]

Harvard University was another locale where sociology and anthropology connected in the late 1930s. Not in its collegiate departments: sociology was decidedly unsupportive of empirical field study; and anthropology, conjoined with archaeology in the Peabody Museum, was more museological than ethnographic, though it did accommodate mavericks like Conrad Arensberg (PhD 1934). Raised in a cultured middle-class background, Arensberg had from an early age aspired to be a writer and chose anthropology as a graduate field because as the "psychological history of mankind" it promised to be good training for a writer. So he happened to be on the spot when the chairman of the joint department, the archaeologist A. M. Tozzer, allocated a departmental research fund to a five-year archaeological survey of Ireland, which included as an add-on an intensive ethnography of one county. County Clare was selected, for its mix of Gaelic and English languages and cultures, and Arensberg spent two and a half years in residence, observing a traditional rural society in transition to a modern economy and lifestyle. A stock museum survey thus became the vehicle for one of the first (and best) ethnographies of acculturation in a Western community.[35]

The most active locus of ethnographic sociology at Harvard, however, was Elton Mayo's Department of Industrial Research in the Harvard Business School, whose main undertaking was the famous Hawthorne study of industrial labor. This extradepartmental arrangement, typical of Harvard, enabled charismatic individuals with access to extramural funding to pursue personal visions unhampered by disciplinary customs and constraints. Lloyd Warner's massive survey of "Yankee City" (Newburyport, Massachusetts) was organized in Mayo's program, with funding from the Rockefeller family. The most ambitious community study since Middletown, it was a model and inspiration for other such projects, including William Whyte's.

How Mayo, an Australian clinical psychologist, got involved in the ethnography of workplace communities is well studied and can be skimmed over here.[36] Like many clinical social workers at the time, he was aware that group behavior—covert worker control of the pace of machine assembly, in the Hawthorne case—could not be understood by either time-and-motion study or individual psychology, but only by inside knowledge of the situations of workers' lives outside the factory. Mayo's low opinion of sociologists cut him off from that source of expertise. ("The sociologists of today in universities are the most incredible asses," he once observed, "who know nothing of the world about them, and take no responsibility for others.") His ideal was a young anthropologist fresh from research in some South Seas island culture who aspired to apply his ethnographic know-how to a modern community; and in 1930 just such a person turned up.[37]

Lloyd Warner, a newly appointed instructor in anthropology, had just returned from a two-year residence in a tribe of Australian aborigines and was looking to apply ethnographic methods to his own society. In fact, he had gone to Australia with just that plan in mind. So Mayo sent him out to South Chicago to talk with officials at Western Electric about carrying out an ethnography of Hawthorne workers in the workplace and, more important, in their homes and neighborhoods. Ethnography was the same, Mayo assured company officials, "whether you study the Australian totemic clans, the gangs of Chicago, or our departmental groups." However, Warner concluded that Cicero, Illinois, where most of the Hawthorne workers lived, was just too ethnically mixed and "disorganized" by vice and crime to qualify for ethnographic study (an oddly conservative view).[38]

Instead, Warner selected Newburyport, a small coastal Massachusetts town whose homogeneous culture and venerable Yankee institutions seemed to him more amenable to ethnographic methods. With a team of field-workers and ample funding, Warner spent the next two years in residence, doing a comprehensive sociological-ethnographic survey modeled on the Lynds' Middletown survey.[39] With an eye to comparative anthropology, he then organized a second and equally ambitious project, staffed by veterans of "Yankee City," of a southern, racially divided town

(Natchez, Tennessee), which in turn inspired Drake and Cayton's *Black Metropolis*. Warner also got Arensberg's Irish study started, carrying out the initial survey and selection of County Clare.[40] Arensberg in turn helped William Whyte get his project under way. Mayo's circle was thus for a time a nursery of ethnographic community studies.

IN "CORNERVILLE"

Whyte was neither sociologist nor anthropologist when he joined Harvard's Society of Fellows as a junior fellow. An eclectic education and omnivorous interests left him uncertain what exactly he wanted to do with his life. The one thing he did know for sure was that he wanted to live among people unlike his own middle-class self and to experience and write about their lives. As an undergraduate at Swarthmore College, he had been attracted to the social and human sciences, but to no one of them in particular. He thought to major in political science but found little about real people and real life in its theoretical abstractions. In the absence of a department of sociology or anthropology, Whyte chose economics, which he hoped would at least touch on issues of economic inequality. Swarthmore's extraordinary alumni network gave him ready access to elite careers in academics, government service, newspaper reporting, law, and public policy. Whyte, however, wanted to finish the novel he had begun to write; and for that Harvard's Society of Fellows was more attractive than any academic discipline. Indeed, junior fellows were barred from joining a department and taking a disciplinary degree.[41]

Whyte had in fact displayed a precocious knack for investigative reporting. As a senior in Bronxville High School (and already a seasoned writer), Whyte had been invited by the editor of a local newspaper to write a column on school issues, in preparation for which he visited schoolrooms to observe and talk to teachers and students about their experience of progressive education. The arrangement was unorthodox, but teachers liked it, and the school's superintendent thought Whyte's reports quite the best he had read on what he was trying to achieve. Whyte did the same while traveling in Germany with his family (his father was a teacher of German), observing firsthand Hitler's growing power. A former teacher thought his weekly columns were more informative than anything in the public media. In 1932, between high school and college, he read Lincoln Steffens's autobiography and was thrilled to discover a man who had grown up in a privileged environment yet was able to engage freely with people different from himself—to understand, not judge or moralize. One is reminded once more of Robert Park. It was a worldly ambition, more than any scientific aspiration, that brought

Whyte into the orbit of the Society of Fellows' genius loci, the physiologist Lawrence J. Henderson, and the Mayo-Warner circle.[42] "The junior fellowship," he recalled, "would free me to do what I really wanted to do: a close-up study of a slum district and let me get to know people and problems unfamiliar in my upper-middle-class environment."[43] More a journalist than a social scientist, he wanted to experience and to write inside stories.

At Harvard Whyte took courses that promised to be useful to his writing project: two on community study, in anthropology; and one in sociology on slum housing. In the latter he carried out a field exercise in the North End, knocking on doors and asking questions about living conditions; but the exercise only demonstrated the ineffectiveness of commuting and questionnaire in getting inside stories. None of his courses offered any practical instruction in connecting with people unlike himself. Nor was the sociology that he read on his own of much help. There was much on "slums" as areas of dysfunctional social "pathology," but nothing showing them as functioning communities, as the Lynds did for middle-class Muncie, and as the social historian and New Deal activist Caroline Ware did for raffish Greenwich Village—an account Whyte particularly admired.[44] Social anthropology seemed closer to what he wanted to do, but its stock monographs on primitive peoples were no help for a would-be ethnographer of contemporary urban communities. A courtesy call on the high theorist and head of the sociology department, Pitirim Sorokin, resulted only in his receiving a lecture on the futility of community study. He had not expected much.[45] Vested departmental interests thus impeded the kind of mixed anthropological and sociological study that Whyte envisioned.

Whyte discovered what Robert Redfield had before him, that operating across disciplines was best achieved at a personal level. Most useful were the friends he made of other junior fellows, especially Conrad Arensberg and the newly fledged anthropologist Eliot Chapple. Arensberg especially had what Whyte most wanted: firsthand experience of living among and observing modern people unlike himself. Arensberg became Whyte's unofficial guide and mentor, discussing methods and concepts, helping to plan the Cornerville project, and encouraging and advising his fieldwork at every turn.[46] (Chapple was more devoted to sociometric methodology.) Whyte thus had access to social-science practices without the constraining pedagogical duties and tribal dos and don'ts of regular departments. With Arensberg as guide, he was free to enter into the lives of others by whatever means worked.

Whyte's initial and very grand plan for Boston's North End was modeled on *Middletown* and *Yankee City*: a comprehensive survey of major communal activities—housing, family, employment, politics, crime, education, recreation, religion, public health, social attitudes. (He envisioned himself directing a team of assistants, but was persuaded by Henderson to begin more realistically, alone.) Getting access to close-knit North Enders was the problem, he quickly discovered. Knocking on doors and direct questioning only made himself and his would-be subjects uncomfortable.[47] Nor were staffers of city agencies and a local settlement house any help. As professionals with middle-class values and biases, they looked down on most North Enders and connected only with those who had rejected their community and were striving to escape it, making them poor guides to local ways of life.[48] Making contact with residents on his own did not work either. An attempt to chat up girls in a bar in nearby Scollay Square as a prelude to getting their life histories only got him an invitation by the girls' male companion to be thrown downstairs. He later learned that no North Ender would dare cross Congress Street to that tough and hostile social terrain.[49]

The way into the Cornerville community, it turned out, was by personal contact and serendipity. Through a settlement-house social worker, Whyte was introduced to Ernest Pecci—"Doc" in *Street Corner Society*—who was the leader of a corner "gang" of young men of about Whyte's own age (he was then twenty-three). Hearing out Whyte's complicated account of his project, Pecci responded that he could easily arrange to get him whatever he wanted to know about the corner group. If it was facts Whyte wanted, he would put questions to the gang, and Whyte would listen to their answers; if it was philosophies of life, he would start an argument among the men, in which their views would be unguardedly expressed. Create the right situations, and normal behavior would do the rest: no problem. Pecci could get Whyte into gambling dens and other local haunts, where strangers could not go. There was nothing Whyte had to do or say to gain access. So long as he was known as Pecci's friend, he would be treated with respect and in time accepted as one of the boys. After his failures on his own, Whyte could hardly believe it would be that easy, but so it proved to be.[50]

The "gang" to which Whyte was introduced in February 1937—the Nortons in Whyte's Cornerville, the Bennetts in real life—was a clique of a dozen or so men who had been friends from childhood. Some were intermittently employed, and some married, but they all still hung out together in Chichi's gambling den or in a tavern, or on a corner of North Bennett Street if there was nothing better to do. The group had been recently organized by Doc after he lost his job with a stained-glass company.[51] Whyte's first appearance with Doc, at Chichi's, provoked (when he went out to use the gents) spirited questioning in Italian of who and what he was (a G-man, maybe?). But as he became known, he could go there by himself and be greeted as an insider, as Doc had said. He also began to hang out on the street corner when Doc was not to be found, and within a few weeks was virtually one of the gang—"just as much of a fixture around this street corner as that lamppost," Doc quipped. North End residents were puzzled at first but quickly concluded that the stranger among them was there to write a friendly book about Italian-Americans. To Whyte's middle-class mind, that account of himself seemed too vague to win his subjects' trust; but to North Enders it was personal relations that mattered. If Whyte was personally all right, his project was all right too: no need to go into particulars.[52]

Whyte did not at first reside in Cornerville but commuted daily from his comfortable student quarters at Harvard. As he became a street-corner regular, however, he discovered that hanging out was a full-time job best done in residence. Finding a room in the overcrowded North End was almost impossible, but the editor of a local Italian newspaper introduced him to the Orlandi family, assuring them that Whyte was not there to "criticize our people," and then persuaded the elder Orlandi son to give Whyte his own room above the family's restaurant, the Capri. Gradually Whyte became like an additional Orlandi son, dining regularly at the Capri, taking part in Sunday family dinners, relaxing at home after a day's work on the corner, and being worried over by a second mother when overdue returning home at night. In 1938, newly married, he moved with his wife into their own apartment (with an actual bathtub), thus becoming a resident and participant observer of Cornerville life.

Not that he was expected to *be* a North Ender: quite the contrary. When he tried too hard to acculturate—as when, to be one of the boys, he once let loose with a string of profanities—he was quickly reminded that his friends did not like him to be out of character. North Enders trusted him and liked having him around because he was reliably himself and different in interesting ways from themselves.[53] Pretending to be what he obviously was not could only stir doubts of his sincerity. Whyte's occupation as a resident North Ender was his science. "Where the researcher operates out of a university, just going into the field for a few hours at a time, he can keep his personal social life separate from field activity," Whyte later reflected. "If, on the other hand, the researcher is living for an extended period in the community he is studying, his personal life is inextricably mixed with his research."[54]

RESIDENT SCIENCE

Whyte's first months among the Nortons were, as he later put it, his "intensive course in participant observation."[55] He had initially assumed that interviews would be his chief sources of data, but quickly learned that formal interviews, or even active questioning, were no way to win trust and get inside stories. The lesson was vividly brought home one evening at Chichi's when, listening to a visiting gambler brag of his exploits, Whyte asked if the cops were paid off. The man's jaw dropped. He glared, then vehemently denied that payoffs were made, and changed the subject. Doc later underscored the lesson: direct questions made Cornervillians suspicious and defensive. In their experience the only people who asked direct questions were interlopers—social workers, government officials, undercover cops—who were there to control, punish, or change North Enders' way of life. Acting as if he knew the unspoken rules of conduct was a sign that Bill Whyte belonged. Then barriers dropped, and he had only to listen and observe to have his questions answered, as well as questions that he would not have thought to ask. "When I had established my position on the street corner," he later recalled, "the data simply came to me without very active efforts on my part."[56] Just being there, observing and taking part in residents' daily activities—in short, residing—was the best way of gathering facts and generating questions and hypotheses. Resident sociology in Boston's North End was thus much the same as social anthropology in Trobriand Kiriwina.

Observed patterns of communal life constituted an interpretive frame derived not from universalizing social theory or preset analytic categories but from residents' own words and actions. Whereas social workers or police saw residents as clients or defendants or other generic social types, resident observers understood their neighbors as they understood themselves.[57] Whyte learned to navigate his adopted community as residents did: by taking part appropriately in the situations of shared lives. His special role was to turn everyday practices of life into practices of science. In a sequestering lab or office it is hard not to see everyday life and science as categorically distinct; in residence it is hard not to see them as a continuum.

It was not just observers whose views were thus transformed; it could also happen to the observed, and did with Doc. At first Doc was simply Bill Whyte's resident guide and informant. but that relation developed into a more equal and reciprocal one. Whyte discussed with Doc what he was trying to do and what puzzled him, and Doc recounted events that Whyte had not witnessed but that Doc knew would interest him, "so that Doc became, in a very real sense, a collaborator in the research." Doc discovered, however, that observing and thinking like a sociologist interfered with his own activities as leader of the Nortons. Rather than acting instinctively and instantly to situations, he paused to think about what Bill Whyte would want to know about them and how he, Doc, would explain them. An observer's self-consciousness slowed the quick social instincts that were vital to a leader's position.[58] As practices of everyday life could become science, so could scientific practices inflect those of everyday life.

This mingling of life and science was a bit unsettling for Whyte the aspiring social scientist. Just hanging out and watching and joining in didn't feel like research, but like relaxing and enjoying himself with friends—the life that began when the day's scientific work was done.[59] A proper scientist would be active and in charge: setting questionnaires, arranging experiments, and sorting data into variables with an eye to statistical proofs of cause and effect. These ideals of scientific methodology pervaded Whyte's Harvard milieu, where Lawrence Henderson enforced a stringent regimen of inductive science on his flock of young social scientists, who in turn strived to embrace practices that would survive Henderson's penetrating critiques. Chapple and Arensberg devised procedures for random-sample observing, and Chapple invented a machine—the "interaction chronometer"—for counting individuals' social encounters, which he believed would establish group structures objectively in terms of a single rigorously defined variable measured mechanically with no taint of individual discretion. Whyte resented Henderson's critiques of his plans for community study, yet these only sharpened his determination to do empirical work that would stand up to anything that Henderson might say against it.[60] We are reminded of Malinowski's brief desire to do his mentor Rivers one better with his own "method of objective documentation."

Whyte was never tempted to give up the intensive personal observing that afforded such an abundance of facts and interpretive ideas. Participant observing, he discovered, fed upon itself: accumulating factual knowledge was punctuated by interpretive insight that in turn opened new lines of empirical investigation. "I had the impression that I was peeling off successive layers of Cornerville life," Whyte recalled.[61] Or as Everett Hughes put it, "The problem of learning to be a field observer is like the problem of learning to live in society. It is the problem of making enough good guesses from previous experience so that one can get into a social situation in which to get more knowledge and experience to enable him to make more good guesses to get into a better situation, ad infinitum." The practices of resident observing developed with use, like those of everyday life. To quote Hughes once more: the participant observer "is caught up in the very web of social interaction which he observes, analyzes, and reports."[62]

Whyte was practicing participant observing before he knew what it was and what exactly he was doing in the North End. That, it turned out, was not a full community survey à la Middletown or Yankee City, but an intensive sociology of the dynamics of small groups. It was not just a difference of scale but a fundamental change of subject, from the institutions and customs of community life to the psychosocial dynamics of face-to-face relations. Whyte thus recapitulated the experience of ethnographers like Malinowski and sociologists like Nels Anderson and Albert Blumenthal, who had gone afield to do large-scale surveys and discovered a small-scale science of everyday life. Small groups proved to be the subjects in sociology for which participant observing was best suited. Subject and method constituted the two halves of resident science.

It was not cool reasoning that caused Whyte to rethink his project, but the press of time. More than halfway into his three-year project, Whyte was still expecting to carry out a full-scale survey of the North End. However, the perspective of a two-month summer vacation in 1937 brought home to him just how much more empirical work he would need to do on important activities—family, religion and the church, politics, economics, the numbers rackets, policing and the courts—on which he had little or no data. Whyte had learned a great deal about the people and the life of his corner gangs, but did not know how these groups might figure in a full community study. "I had the feeling I was doing something important," he recalled, "but I had yet to explain to myself what it was."[63]

The solution to his problem came to him in the fall of 1938. His fellowship would expire the following summer, and to get a renewal—he hoped for another three years—he must show enough progress to justify it. Writing up his material on the Nortons and a rival gang as case studies was a small start but would at least demonstrate that he had not been frittering away his time hanging out with friends.[64] With this thought in mind, he revisited his data files, which it happened he had organized not in sociological categories but by corner groups. That had been a practical decision of data management. As facts piled up, Whyte had to arrange them in some way for ready retrieval; and not knowing what categories of community life would figure in his survey, he filed the data in the easiest way, by corner groups, thinking he could re-sort them into proper analytical categories when he knew what he doing.

In writing up his case studies, however, Whyte realized that the groups themselves were the natural units of social analysis, not the institutional and occupational categories of community studies. Re-sorting all statements about racketeers under "racketeer," for example, added little to his knowledge of racketeering; but keeping them sorted by groups revealed much about the differing attitudes and social situations of his rival gangs. An arbitrary category of data management thus became a category of sociological reasoning. Filing revealed that corner groups were his subject, though he hadn't known it. And his participation in his friends' lives was not a prologue to science but the science itself. These were timely revelations. Whyte was granted only a one-year extension of his fellowship: too little time for a community survey, but enough to complete a sociology—or ethnography—of street-corner society.

BOWLING TOGETHER

As Malinowski and by then many others had discovered, what resident observing was really good for was getting at the psychosocial codes of everyday life. These structures and habits were not revealed by survey interviews and questionnaires: those were designed to give the official view—the *ought* not the *is*. The functioning principles of group life had to be inferred from the small clues of unself-conscious words and deeds—Malinowski's "imponderabilia"—and these were gained only by being inside and taking part. As Whyte wrote, "Only as a participant would I have been able to associate closely enough with these men to work out the structure of the group. If my information had been limited to personal interviews, this would not have been possible. When I asked one or another of them who their leader was—as I did from time to time—the answer was always the same: 'We have no leader. We are all equal.'"[65]

Despite their firm denials, the Nortons obviously did have a leader—Doc. What's more, Whyte discovered, they had a status hierarchy in which every member of the group had a definite place, of which he and everyone else were aware. Those at the top—Doc and his underleaders—were the ones who could initiate group activities; those at the bottom followed their lead. That was the master principle of group order, and the order of standing was maintained and adjusted by repertoires of behavior that everyone understood and manipulated, yet denied if directly questioned. There were no rules; what people did was just what they had always done: that was the official reality in Cornerville, as it was in Omarakana or the neighborhoods of Chicago's Near North.

The behavior of leaders was especially vital to group order, because leaders were expected to enact most punctiliously the group's unstated rules of social standing and authority. They were leaders precisely because they were the most perceptive of how the group worked and the most adept in manipulating situations and individuals' behavior to maintain group solidarity and order. That was why Doc could so confidently promise Whyte that he could get him anything he wanted to know by creating situations that elicited the desired behaviors. It was what, as the group's leader, he did all the time.[66] One reason why Whyte had not seen that the group as a whole had a structure was that the activities through which that structure was created and maintained were recreational activities like baseball or bowling, activities that for Whyte were the least scientific and the most recreational.

The event that opened Whyte's eyes occurred in April 1938 in the course of a competitive bowling match among the Nortons. It was, he recalled, "one of the most exciting research experiences in Cornerville."[67] Bowling had become more than a casual recreation among the Nortons in the fall of 1937, when Doc organized a competitive match with the Italian Community Club, a rival group based in the local settlement house and composed of college graduates or college-bound men who, the Nortons felt, disdained them as unambitious dead-end kids. The purpose of the match was to put the middle-class wannabes in their place—which in the event, they did. A side effect of that triumph was that bowling became an important means of establishing standing within the Nortons themselves, in weekly matches between teams drawn up by Doc and his underleaders. These matches were obsessively discussed before and after: who would play well and who would not; who had deserved to win and who had just been lucky or played over their heads.

Before the April 1938 match, the season's grand finale, Doc and his inner circle predicted a rank order of performance, which as it happened closely followed the group's rank order of leaders and followers. Whyte at first gave little thought to their prediction, which was his as well. He had previously noticed, without giving it much thought, that in baseball games high-ranking members performed best even if they were not the best athletes, while some of the best athletes among the lower-ranking Nortons performed poorly. It was the same pattern with the bowling match. This time, however, the result had been publicly predicted, and the prediction fulfilled: the higher a man's standing in the group the better he bowled. Two exceptions only proved the rule: Whyte himself, and a man who bowled with the Nortons but was not a member of the gang. Both performed above their usual level, and both were outsiders whose standing in the group did not matter. It was a natural experiment in which a hypothesis was made and empirically tested, with controls. It was perhaps that likeness to science that got Whyte thinking about causes and effects: what social mechanisms linked social standing to performance in a game? Whyte felt uneasily certain that he had witnessed the enactment of an important social principle, though exactly what it was he could not yet see. It was not until he was reviewing his observations of the context and events of the bowling match that the lightbulb went off.[68]

The psychosocial mechanisms that linked performance to social standing had been in plain sight all along, Whyte realized. Sports and other performative activities were always accompanied by running commentaries on individuals' abilities and performances, and it was this verbal support and disparagement that brought performance into line with social standing. In skilled sports especially, performance depends on self-confidence; and confidence can be manipulated by peer behavior. When lower-status members of the group were bowling unbefittingly well, their confidence was undermined by persistent and often vicious heckling—they were just lucky, or performing over their heads—with the result that their performance collapsed. It wasn't the leaders who did this cutting down to size, but usually men of middling status striving to protect their positions from strivers just below them, or to edge up from follower to underleader. The most vulnerable were the most aggressive. Doc and his underleaders consciously manipulated this social dynamic. It just wasn't right for men to perform better than their standing in the group, they told Whyte when he asked about it. It would only cause arguments and require another match to set things right.[69]

Whyte realized to his chagrin that he had been missing a unique opportunity to document with quantitative precision the behaviors that made and remade social structure. A numerical record of every string bowled by every one of the Nortons for a whole season would have made him the envy of his sociometric friends Arensberg and Chapple. Yet he had seen no reason to record as science what he was experiencing as recreation. He had in fact felt guilty for spending so much time having fun instead of sticking to his scientific work of gathering data. When all the time the recreation *was* the science:

> I was bowling with the men in order to establish a social position that would enable me to interview them and observe important things. But what were these important things? Only after I passed up this statistical gold mine did I suddenly realize that the behavior of the men in the regular bowling-alley sessions was the perfect example of what I should be observing. Instead of bowling in order to be able to observe something else, I should have been bowling in order to observe bowling. I learned then that the day-to-day routine activities of these men constituted the basic data of my study.[70]

Whyte thus rediscovered Malinowski's "imponderabilia of everyday life." Once again, an unknowing participant observer became one knowingly.

Whyte in fact experienced the mechanism of social ranking on his own person in the big match. As Doc's friend he had a high social standing and was thus expected to excel, and his friends made sure that he did:

> I simply felt myself buoyed up by the situation. I felt my friends were for me, had confidence in me, wanted me to bowl well. As my turn came and I stepped up to bowl, I felt supremely confident that I was going to hit the pins

> that I was aiming at. I have never felt quite that way before—or since. Here at the bowling alley I was experiencing subjectively the impact of the group structure upon the individual. It was a strange feeling, as if something larger than myself was controlling the ball.[71]

His own visceral experience thus confirmed his scientific understanding of what he had observed. He had felt the emotions that brought physical performance in line with group standing. That understanding was vividly reinforced when Doc, piqued by his friend's incautious (and unjustified) gloating over his bowling prowess, arranged a rematch with the lowest of the leaders, Long John, who, cheered on by the Doc and others while Whyte was heckled as bowling over his head, beat Whyte handily. Put in his place socially, Whyte settled for a sociological discovery.[72] It was participant observing it its most powerful form—*participant feeling*, one might say. It was a little social experiment, with Whyte as both subject and observer.

Other events revealed that the same psychosocial mechanism also connected social standing and mental health. The first such episode followed Doc's partial withdrawal from the Nortons when he began hanging out with a clique led by a gambler, Spongi. Doc's absence triggered a reshuffling of the group's hierarchy, as followers began to attack underleaders in the hope of moving up to leadership positions. As the lowest underleader, Long John was especially vulnerable, so was pointedly ignored when at Spongi's and subjected to unrelenting heckling by Nutsy, his chief rival, and Nutsy's allies. Whyte had noticed that Long John was bowling poorly and was hanging out less often with Doc, but he didn't see the connection until Doc happened to mention that Long John was suffering from depression and a recurrence of childhood nightmares. Then everything snapped into place: "the three-way connection between group position, performance, and mental health." To test this hypothesis—and to help a friend—Whyte suggested to Doc that he restore Long John's standing by arranging positive and visible social interactions. Doc was skeptical but agreed to try. He began to encourage Long John in bowling and to include him in every activity at Spongi's, treating him ostentatiously as a favored member of the group. The experiment worked: as Long John's social interactions were restored, his symptoms disappeared, and he became a prize-winning bowler. The good deed was, Whyte recalled, "a nice bonus in the development of interaction theory."[73] Science became life, as life had become science.

A similar scenario played out with Doc himself, when for lack of pocket money he could no longer spend freely as a leader was required by his status to do, so withdrew from social interactions and as a result began to suffer dizzy spells whenever he was in the group. Doc understood the cause of his distress but lacked the means to cure it, with a paying job. Jobs for corner boys were very hard to get. But as an educated Harvard man, Whyte had the social standing in the middle-class world to do what Doc could not. He persuaded the social workers in the local settlement house to hire Doc as a project manager, a post they would ordinarily never have given to a corner boy. As Whyte expected, Doc was a great success, thanks to his deep knowledge of the settlement's clientele—knowledge that the socially insulated social workers did not have. As Whyte had predicted, Doc's dizzy spells soon disappeared.[74]

Whyte thus showed that resident science could go beyond observing to experiments, in which observers and subjects between them had sufficient knowledge to set up situations that paired conjectured causes and effects. In the experiment with Long John, Whyte had the scientific knowledge to connect cause and effect but lacked the social position and know-how to set up the experiment. Doc, in contrast, had the standing and know-how, but lacked the scientific conviction that Long John's mental health could be deliberately restored—even though it was just the kind of social manipulation that as the Nortons' leader he had unthinkingly enacted countless times. The experiment with Doc was a variant of that: here it was Whyte who had the social standing to set it up, and Doc who had the know-how to carry it out. These joint interventions transgressed the scientific requirement that experiments be objectively hands-off; yet they were no less effective experiments for that. When participant observation becomes participant intervention, the ideal of detached objectivity gives way to what the sociologist Allan Holmberg has suggestively called " 'objective' subjectivity." Whereas laboratory experiments achieve control by losing context, experiments in situ gain access to the "total situation" of real life by limiting formal control.[75] Experiments are best guided by epistemologies that suit their particular situations: one size does not fit all.

Although Whyte worked most intensively with his corner groups, he did not entirely neglect the community institutions that constituted the groups' social world. Inside observing of all these activities was obviously out of the question, so he confined himself to the two that mattered most to the Nortons' daily lives: namely, the numbers rackets and local machine politics. These were the institutions that enabled North Enders to take part in economic and civic life, legitimate business and city politics being largely off-limits to them. They were also much harder for Whyte to access. In the larger, darker worlds of racketeering and machine politics, Doc could not open doors and create situations. Methods that worked effortlessly with the Nortons did not with work at all with politicians, cops, and numbers runners. So Whyte turned instead to the detached techniques of camouflage and watching unobserved from the sidelines.

To get access to politicians, Whyte took a job as volunteer assistant to the secretary of a state senator allied with a mayoral candidate who, though flamboyantly corrupt, was known in Cornerville as a friend to Italian-Americans. In this role Whyte could sit unobtrusively in the senator's office in a local funeral parlor, observing and, in the guise of taking minutes of meetings, recording how politics worked at the grassroots. It worked much the same as corner groups, he discovered, as a system of reciprocal favors and obligations, with corner-group leaders serving as links between citywide system and street. He largely avoided the entanglements of actual participation in political machinations. As an outsider he had no votes or endorsements to deliver, so could not play in that game. He did once allow himself to be persuaded to repeat vote (a common local practice), which nearly got him arrested and taught him that the hazards of participation were not just safely epistemic but also moral and legal.[76] But he learned enough from his undercover makeshifts to sketch in the connections of corner groups to the political system, though not with the deep understanding that came from sharing the lives of his corner groups.

It was the same with the numbers rackets. Here again Whyte sought to gain access by establishing a friendly relation with an insider: a racketeer friend of a friend, who he hoped would afford an inside view of a world closed to outsiders. The relation soon cooled, however, probably when "Tony" realized that Whyte had nothing to give him in the way of useful connections to the middle-class world outside Cornerville, which was what he had expected as a quid pro quo for access.[77] Rackets, like politics, operated on the principle of reciprocity; and Whyte lacked the social currency to reciprocate. Unable to engage and observe on the inside, he fell back on the hands-off practices of the "new social survey," like Eliot Chapple's sociometric technique of counting and cross-tabulating individuals' social encounters to map group structure. He watched and counted interactions in a group of racketeers through a window of their hangout, which was fortuitously just across the street from the Whytes' apartment and so could be monitored unseen. In this way Whyte got enough data to identify leaders and followers in the group, and to understand in a general way how racketeering connected to corner groups through their leaders.[78]

Whyte's truncated studies of racketeers and politicians reveal the limits of participant observing in a contemporary society. The scale and social locations of economic and political institutions were just not amenable to participant observing, and their social customs precluded direct access by strangers. So Whyte remained an outsider looking in, much like a social psychologist in a laboratory observing his subjects through a one-way glass and recording atomistic data in predefined categories on machines or printed forms. Check, check, check, check.

Whyte's small-group sociology was in several ways unlike the community sociology out of which it unexpectedly evolved. Community study was survey science; small-group sociology was a science of particulars.[79] The one tended to Galilean abstraction, the other to Sherlockian clues and conjectures. Actors in large community studies were generally depicted as social types; Whyte's actors were individuals with names (or pseudonyms) and life stories. In community surveys, participant observing was an embellishment. In studies of small groups, it was the chief means of discovery, and the operating principles of group life were set forth not in the universal laws to which survey science aspired, but in stories of particular actors, actions, and events. Survey sociologists reasoned abstractly with generic data; small-group sociologists reasoned concretely with particulars in context. As Whyte recalled, "The ideas grow up in part out of our immersion in the data and out of the whole process of living."[80] Community survey described a current state, while Whyte depicted groups in continual change. He was, as he put it, "taking a moving picture instead of a still photograph."[81]

As we experience and understand our own unfolding lives as stories, so do resident observers experience their science as entering into stories in progress. Herbert Gans recalled that as a participant observer in Levittown he often felt like "a vicarious participant in a large number of dramatic serials, some involving heroes fighting villains, others with cliff-hanger endings, and as soon as one serial came to a dramatic (or an undramatic) conclusion, others were sure to have begun."[82] Science in situ was often science as sitcom. It was the same with Jane Goodall's study of chimp social behavior, as we will see in the next chapter. Whyte subsequently wrote a novel of life in the North End with Doc as its central character, while he was recovering from a crippling attack of polio in 1943–44. But when an agent confirmed his own suspicion that it was no good, he decided that *Street Corner Society* was in fact the novel he had always wanted to write, about people whose lives were unlike his own.[83]

CONCLUDING THOUGHTS

Tension between Sherlockian and Galilean practices runs deep in the social and human sciences: on one side situating, observing, cases, and stories; on the other side desituating, counting, abstraction, and laws. These modes are not irreconcilable: investigators not uncommonly use both, as Whyte did. But in the post–World War II decades of rapid demographic expansion and professionalization in all the sciences, the difference between them hardened. As polling survey and big-data big-science sociology became a powerful mainstream, small-data individual observing flourished in satellite fields like small-group sociology, industrial and human relations, and a scattering of ethnographic community studies in the style of *Street Corner Society*.[84] High priests of polling tended to dismiss these community studies as unscientific.

The divide between two subcultures is strikingly apparent in the views of the sociologist Samuel A. Stouffer, an early and ardent advocate of the new polling survey. He grudgingly allowed that in situ observing was useful as preliminary exploration; but even that, he feared, was a potential diversion and waste of time:

> It may be that in sociology we will need much more thinking and many more descriptive studies involving random ratlike movements on the part of the researcher before we can even begin to state our problems. . . . However, I think that we can reduce to some extent the waste motion of the exploratory period if we try to act as if we have some a priori ideas and keep our eyes on the possible relevance of data to these ideas. This is easier said than done. So many interesting rabbit tracks are likely to be uncovered in the exploratory stages of research that one is tempted to chase rabbits all over the woods and forget what his initial quarry was.[85]

Given the cost in time and money of exploration, Stouffer continued, "it is obvious that we cannot afford the luxury of conducting them as isolated fact-finding enterprises. Each should seek to be some sort of *experimentum crucis*." He was "not in the least deprecating exploratory work in the field," he hastened to add, just calling for "some orderliness."[86] But his words—"random ratlike movements," "rabbit tracks," "waste motion"—leave no doubt that "order" in his mind meant the reductive and desituating procedures of big-data polling: that, and that alone, was science.

Academic departments divided between humanistic and scientistic approaches. In the Society of Fellows, Whyte had been largely insulated from academic clannishness, but he ran into it head-on when on Arensberg's advice he went to Chicago to study with Lloyd Warner and get his PhD and academic job card. Whyte was open to either sociology or anthropology—Warner had a dual appointment—and chose sociology only because to become an anthropologist he would have had to study physical anthropology and archaeology, which interested him not at all. He then discovered that Chicago sociologists were divided between a faction (Warner and Everett Hughes) who favored qualitative, in situ fieldwork and accepted their subjects on their own terms; and another (Louis Wirth, Herbert Blumer) who favored high theory and defined "slums" as pathologically disorganized. Students of Hughes and Warner, Whyte recalled, had to watch their backs. To turn his Cornerville manuscript into a properly scientific dissertation, he was pressed to trim the narrative elements (the book's great strength) and add the footnotes and literature review that were the obligatory signs of proper science. He declined. Nor did he yield on his conviction that the operating principles of community life had to be understood in subjects' own terms, not those of social theorists. As he put it, "The middle-class person looks upon the slum district as a formidable mass of confusion, a social chaos. The insider finds in Cornerville a highly organized and integrated social system."[87] Thanks to Hughes's intervention Whyte was, despite his stubborn heresies, duly credentialed as a sociologist.[88] However, he made his career not in mainstream sociology but in the more pragmatically open field of industrial or human relations, in which inside observing of small groups was the chief and defining practice.

Reflecting in 1968 on the state of his discipline, Herbert Gans wondered what drew some sociologists to resident observing, even as most rushed pell-mell into large-scale team research and computer-aided number crunching. Perhaps, he conjectured, it was individuals who were alienated from the societies in which they were brought up who were strongly drawn to experience the lives of people unlike themselves.[89] It seems to me more likely, however, that individuals who were drawn to ethnographic sociology were not alienated by their own social experience but rather inspired by it to experience other ways of life as well, and to discover from the inside what made each one work. Sociologists became participant observers not to leave their own lives behind but to live them with a broader and deeper understanding. That is another conjecture, to be sure. But it is supported by the life stories of the people I've discussed in this book, and others I have not discussed but would have liked to—the anthropologist Hortense Powdermaker, Herbert Gans himself. In their lives there is little evidence of alienation and much of eager engagement with the endless variety of human social life.[90] Gans called for a systematic prosopography of resident observers in sociology—an excellent suggestion that remains for someone to take up. I would only add that such a study should include ethnographers, and resident observers in all the social and life sciences—or any science. I like to think that the case studies in this book are a modest move in that direction.

CHAPTER 5

IN CHIMPLAND

JANE GOODALL

Of all the principal characters in this book, Jane Goodall may be the only one who needs no introduction: she is one of the most familiar and celebrated scientists of our day. A media sensation almost from the start of her work at Gombe, she remains newsworthy as an activist for chimpanzee and animal welfare. Her celebrity derives not just from the originality and excellence of her science, but also in part from her life story: a spunky and courageous young unknown, untrained in science, who lived in the East African forest among powerful and volatile animals, and by personal qualities of passion, intuition, and empathy for chimps overturned our view not just of them but of ourselves. Me Jane, no Tarzan—it was a story for its time. The iconic "Jane Goodall" is not my subject here, however, but the exacting and imaginative observer of animals in nature: the Jane Goodall who turned abilities developed in her personal life into a science of resident observing; the Jane Goodall of subtle clues and bold conjectures.[1]

The real Jane Goodall was not just gifted but also lucky in her timing. She was in the Gombe forest inventing her distinctive scientific self and practice in that "brief sliver of time" in the early 1960s when primatology was being reinvented as a field science after a pause of nearly twenty years.[2] A tabulation of person-months of fieldwork on primates from 1930 to 1966 reveals the dramatic upturn that occurred about the time Goodall arrived at Gombe. For over twenty years of depression, world war, and recovery, almost no fieldwork was done; then some in the early 1950s; and from the late 1950s, a rapid and sustained growth doubling every five years. Most of this fieldwork was carried out on old-world monkeys, with work on apes (gorillas, chimps, gibbons, orangutans) constituting about one-fifth of the total by 1965. In the 1960s there were six substantial field studies underway of chimps in nature: four short-term (two to nine months), and two of many years: Goodall's in Gombe, and a group project of Japanese primatologists led by the founder of the science in Japan, Kinji Imanishi, and his student and successor, Junichiro Itani.[3] Goodall went afield in 1960, at the precise moment when study of apes in nature was becoming practicable, not just for professional ethologists but for others who were just in the right place at the right time. Primatology was a wide-open field in which foundational practices could be defined or redefined. Just a few years earlier, there would have been few opportunities for someone like Goodall to do what she did; a few years later, someone with scientific credentials would likely have been the "Jane (or the Jay or James) Goodall" of chimps.

Primatology in the years of its rebirth was exceptionally varied in its practitioners and practices. Primates, because they are close biologically to humankind, attracted followers from a range of behavioral sciences—anthropology, psychiatry, psychology, physiology, sociology, zoology, ethology—each with it its own trademark practices and shibboleths. Anthropologists, with their strong tradition of participant observing, favored intensive observing of free-living animals in situ, whereas those from the laboratory sciences favored quantitative and quasi-experimental approaches that were done outdoors but not in natural situations. Psychologists, one observer quipped, regarded the field situation as "an immensely disorganized and uncontrolled comparative psychology laboratory." Zoologists and ethologists approached primates as they did birds or small mammals, gathering aggregate data on typical behaviors by survey-style methods, rather than facts of particular individuals and situations.[4] Untrained in any science, Goodall developed a practice of close following and observing—"habituation"—that was none of the above but resembled ethnography or small-group sociology in its concern with individuals and contexts of ordinary life.

Goodall was not the only one to pursue a practice of walking and watching. George and Kay Schaller, in their work on mountain gorillas in Congo and Uganda, were early practitioners of habituation.[5] So were Japanese primatologists like Imanishi, who, like the Japanese generally, assumed a continuity between animals and humans that Europeans did not; they worked first with commensal Japanese macaques and then with African apes.[6] Vernon and Frances Reynolds tried habituation with chimps in the dense Budongo Forest in Uganda, though they made little headway in that forbidding environment (hard to get around, not much to see).[7] But once the Schallers and Goodall demonstrated what could be achieved with habituation, many others followed: most famously Dian Fossey with the gorillas of the Virunga Mountains in East Africa, and Biruté Galdikas with the orangutans of Indonesian Borneo.[8]

Goodall's resident observing owed little to the practices of other ethologists, however, because she was largely unaware of them. What did matter was her life experience before Gombe and what she did in her first and formative year of forest walking and watching. The history of primatology is thus a less appropriate historical frame for Goodall's science than the analogous experiences of resident observers in the human and social sciences like Bronislaw Malinowski, Nels Anderson, and William Whyte. My subject is how resident practices develop in the varied contexts of different disciplines, subjects, and individual lives, and across the divide between the human and the animal sciences.

GOMBE

Unlike Malinowski at Omarakana, Goodall really did arrive at Gombe on a beach by boat, after a short run along the scalloped shore of Lake Tanganyika from the village of Kigoma some twenty miles to the south. Like Malinowski, she arrived not alone but in company: her mother and chaperone, Vanne; Dominic Bandola, a cook and all-purpose man hired in Kigoma; and the head ranger for the area, David Anstey. They landed at the mouth of a small stream in one of the many bays between rocky headlands and were met by a small crowd of Africans: two resident game scouts with their companions, and some of the fishermen who camped nearby. After a (long) speech of welcome by the honorary "headman," the three new residents followed Anstey a short way through the scrub and forest to a small natural clearing, where they set up a large tent for Jane and Vanne and a smaller one for Dominic. Vanne later admitted to feeling horrified at the impenetrable forest in which Jane would be working, while David Anstey was thinking to himself that this untried young woman would last six weeks at most. Jane herself felt "neither excitement nor trepidation but only a curious sense of detachment," and recalled sensing that this strange forest world was where she was meant to be.[9] It was the primal experience of would-be observers "suddenly set down . . . on a tropical beach" (in Malinowski's words) to live and to learn.

The Gombe Stream Chimpanzee Reserve is a thin lakeshore strip of rugged terrain that forms the escarpment of the East African rift valley.[10] Dissected by the steep ravines of a dozen major streams, it is covered with vegetation that changes from beach and open woodland at lakeshore, through tropical evergreen and deciduous forest with brushy understory in ravine bottoms and on lower slopes, to a patchwork of forest, grassland, and savannah at middle elevations and open grassland near the top. The reserve in 1960 was home to about 160 long-haired chimpanzees (*Pan troglodytes schweinfurthii*) in four communities, each with its own home range. Goodall's community—the Kasakela—ranged over the Linda, Kasakela, Kakombe, and Mkenke valleys, in the north-central quarter of the area. The reserve was also home to baboons, four species of monkeys, and a variety of smaller mammals and snakes; large and dangerous animals—lions, hippos, buffalo, leopards, crocodiles—were a fading memory.

Because the area had been a protected game reserve since Tanganyika became a German colony, and before that a sacred place for Africans, it had no permanent human residents and had never been timbered or farmed, or hunted for bush meat. The Gombe chimps moved freely in habitat that was ecologically largely intact and undisturbed by human clearing or foraging. The rainy season from October to May was miserable, but summers, though hot, were relatively comfortable. And though the place was malarial (as Jane and Vanne soon discovered), the streams were pure and potable, and the lake free of bilharzia and safe for bathing. Regular boat service (for the game wardens) made Gombe accessible for supply, medical care, visitors, and periodic respites, yet not too accessible. Visitors were very few and strictly regulated. Gombe was, in short, a fine place for watching chimps undisturbed in situ—far better than the wild and dangerous rain forests in which most African chimpanzees lived. In the next five years Goodall was in residence for forty-five months: from July 1960 to December 1961, June to December 1962, April to December 1963, and April 1964 to March 1965. A permanent research station with year-round staff was established in 1965.[11]

How Goodall got to Gombe is an oft-told tale. She had dreamed from an early age of living and communicating with animals: a dream inspired by the fantasy worlds of Hugh Lofting's Doctor Dolittle books and Edgar Rice Burroughs's Tarzan series and imperfectly realized in her childhood activities with the small creatures of garden and countryside—frogs, turtles, snakes, spiders, worms, birds, butterflies, small mammals. As a young adult Jane yearned to go to Africa, with its large and dramatic wildlife; so when a school friend, Marie-Claude "Clo" Monge, invited her in 1957 to visit her family farm in Kenya near Nairobi, it was a dream come true. She took up residence in Nairobi, supporting herself as the personal secretary of Louis Leakey at the Coryndon Museum.[12]

Goodall was then entirely untrained in science. Though good in school she had not gone on to university. That was partly for lack of money (her father was absent and sparing in providing for Vanne and two daughters). But it was also because Jane experienced formal education as a road to some routine middle-class job, which was not the life she meant to live. A school adviser was perplexed that such an able student would want to spend her life watching wild animals in Africa, but to Jane the work open to women was not an option. She thought seriously of a career as a journalist—she liked literature and had become a very good writer—but that uncertain occupation was not a practical choice for someone who had to start earning a living right away. Vanne, sensing that her unorthodox and headstrong daughter would have an unpredictable life course, suggested she go to secretarial school: not because she thought Jane would ever be a career secretary, but because secretarial skills would get her a job wherever in the world and in whatever situation she wandered into—a freedom that journalism would definitely not afford.[13] It proved to be sound practical advice. Jane got to Africa and, three years later, to Gombe, not by following a straight and well-worn path of schooling but by her habit of finding work and seizing opportunities that unexpectedly came her way.

The deus ex machina of opportunity for Goodall was the ubiquitous and hyperactive Louis Leakey. Leakey was a famously colorful figure: precociously intelligent (a double first at Cambridge in anthropology and archaeology), unorthodox in his upbringing (as much Kikuyu as colonial British), and ebulliently eccentric. He was physically energetic and interested in everything, a man of enthusiasms; a great talker and persuader, sociable and charismatic, maddening and irresistible—all quite amazing for a son of missionaries.[14] His chief passions were the paleontology of East African Miocene apes, and paleoanthropology—stone tools and their protohuman makers—as well as East African ethnography, and anything having to do with animals living or extinct. As head curator of the Coryndon Museum, he had a diverse social network in both Africa and Britain—he seemed to know everyone—and was the center of a vortex of field projects. In 1958 he helped found (and fund) the Tigoni Primate Research Centre for study of a colony of East African monkeys.[15] Other projects aimed to demonstrate that knowledge of living primates in nature could illuminate the way of life of the protohumans whose tools and remains he and his wife, Mary Leakey, were excavating at Olduvai Gorge. He had sent out a man to watch chimps in Gombe in 1946, and a decade later dispatched his then secretary, Rosalie Osborne, to watch gorillas in the mountain forests of Uganda, which she did until her mother got wind of the adventure and called her home. Goodall's was Leakey's third such project, and would be followed by those of Dian Fossey and Biruté Galdikas.[16]

Leakey was thus a generator of opportunities for talented and often untrained young people, especially young women, who were inexpensive and eager to engage in projects carried out with minimal organization and means. As a man of museum and field, he had a bias against the academically credentialed, and he believed women were more acute and exacting observers than men. As Vanne later observed when Leakey was arranging temporary work for Jane in London, "He's like a magician. Want a job? British Museum? Zoo? Natural History Museum?"[17] Leakey arranged them all—the magician pulling invisible strings and making impossible things happen.

Goodall got wind of Leakey soon after arriving in Nairobi, and without waiting for a proper introduction visited the museum where, having displayed her passion and knowledge of animals and her acceptable typing and shorthand, she was hired on the spot as Leakey's secretary, replacing the departed Rosalie Osborne.[18] A few months later, having proved her enthusiasm for fieldwork and a capacity for hard work in difficult conditions (wrangling a leopard cage through thick bush), she was invited to join the Leakeys' annual summer dig at Olduvai. There she further demonstrated her taste for living rough and for the hard and exacting manual labor of fieldwork. In the last days of the expedition, and then later at the museum, Leakey revealed his project of studying African apes in the area of Africa where *Homo sapiens* evolved. He regaled Goodall with the difficulties and dangers of the work and the extraordinary qualifications that anyone must have who dared attempt it. Not imagining that she might qualify, Goodall finally told Leakey to please stop talking about the project, because it was just what she had always longed to do. Leakey then confessed that he had been hoping she would volunteer, and that the personal qualities of passion and endurance she had demonstrated at Olduvai were all the qualifications she needed. Scientific training in his view only got in the way of unbiased observing. So the odd little dance—Leakey not saying he wanted Goodall to go to Gombe, Goodall not saying she wanted desperately to go—ended in the scheme that after three years and some further setbacks landed her at Gombe.[19]

Goodall would need a female companion—a problem solved when Vanne surprised herself by impulsively volunteering. Funding was a greater problem, solved two years later, in 1959, when an American philanthropist who had been writing modest checks for the work at Olduvai agreed to an additional and larger grant for a four-month trial at Gombe. He also offered to send a large live-in land yacht disguised as an elephant, with a head fabricated by a firm that made props for Hollywood films, a ruse he felt certain would fool the chimps and get observers in close. Leakey politely declined the elephant-yacht, but took the money.[20] And so in July 1961 Goodall arrived at Gombe, aged twenty-six and marveling at her good fortune. "What a life, eh!" she wrote her family a few months later: "Sometimes it suddenly comes over me how strange it all is, really. Here I am, an ordinary person, with my staff of 3, a camp, unlimited funds, and doing what I have always wanted to do. Not stuck away in some horrid office, out of the sunlight, but out in the open, sleeping under the stars, climbing the mountains, watching all the animals. Is it possible? Can it *Really* be me? Or is it some strange hallucination?"[21]

The path to Gombe seems all the more unlikely when set among all the default life paths that Goodall could more easily have taken but did not. In England she could have learned the photographer's trade, as a career counselor advised, and lived by taking endearing photos of people's pets. In Kenya she could have become a horse trainer, or the wife of a horsey polo-playing colonial or a big-game hunter and organizer of safaris—she had plenty of admirers and offers. Meeting a game warden and his wife camped happily in the bush afforded a glimpse of another possible life among animals. A less romantic but more likely scenario was the life of an assistant in a zoo or museum, which Goodall experienced at the Coryndon and later in a brief stint at the London Zoo helping Desmond Morris and his wife produce popular nature films. She liked the variety of museum and zoo work but could not abide the sight of dead and captive animals.[22] None of these possible lives engaged Goodall's taste and talent for observing and thinking things out, which is what—perhaps not so improbably after all—got her to where she most wanted to be.

FOREST WALKING AND WATCHING

What Goodall saw upon arriving at Gombe was not what Leakey had led her to expect (he never did see the place firsthand). Leakey conjured up a forest parkland with chimps living by the lake and in open country, where they could be easily observed. In reality, as Goodall quickly learned, they lived mainly in the dense forest understory of ravines and avoided open places, where there was little for them to eat. Living among them would not be as straightforward as Leakey had imagined.

Observing chimps at home in their own world meant being close yet unobtrusive, so the animals would carry on as if observers were not there at all. One way was to stay hidden in camouflaged blinds—the common method of hunters or ethologists, like the zoologist and early chimp watcher Adriaan Kortlandt. The disadvantage of stealth was that it was how predators stalked their prey, and chimps were quick to spot and flee from what was hidden. Another method was to stay in plain sight, keeping distance at first and slowly getting closer until animals became accustomed to observers' presence and went about their lives as if at home alone. But habituating had the disadvantage of being slow, unpredictable, and low yield when animals didn't show up where expected or were stubbornly shy. On the other extreme was the method of provisioning: putting out food to attract free-living animals and keep them where they could be observed. However, provisioning disturbed normal foraging behavior, and provisioning sites were as much human as animal ground, thus subverting the point of observing animals in residence.[23] The advantage of predictable and efficient production of data by provisioning was thus offset by epistemic doubts of the meaning of the data thus obtained.

Habituation, the most intimately residential of these options, was the one least used by professional ethologists at the time. Yet it was the one that Goodall adopted from her first days in the field, on Lolui Island in Lake Victoria, where Leakey arranged for her and Vanne to spend several weeks observing a colony of vervet monkeys, as a warm-up for Gombe. It was an easy apprenticeship. Moving about as a whole community, the vervets were easy to locate and approach; and when their alarm calls signaled to Jane that she was too close, she curled up and pretended to sleep, thereby putting the animals at ease. Within days she had begun to recognize individuals and was getting data of individual and group behavior.[24]

Habituation proved more difficult with chimps. They did not move about conspicuously in whole troops, but foraged in small groups over large forest territories. Just finding them was a challenge. Goodall had beginner's luck when her two African minders noticed chimp activity at a large msulula tree coming into fruit in the northernmost valley of the reserve. From a vantage on a distant hillside, she could just make out chimps in the dense foliage stuffing themselves, but little more; and when she moved down into the valley to get a closer look, the chimps fled. And when the tree was stripped, the chimps just vanished. Goodall, with her reluctant minders in tow, spent the next month struggling up and down one ridge after another with nothing to show for it but occasional glimpses of animals melting into the bush. Chimps were more visible from the open ravine slopes, but so were observers: spotting them from five hundred yards, the chimps fled into thick undergrowth.[25]

The main problem with habituating chimps stemmed from their habit of living in small and constantly changing groups, a habit that was adapted to the unpredictable ripening of the large fig and msulula trees that provided chimps with their stable foods. Forest wandering was for Goodall likewise an experience of unpredictable bonanzas in a daily routine of exhausting dawn-to-dusk slogging through rough terrain with little to show for it at day's end. The ecologies of foraging for figs and for scientific data were thus much the same. After a month or so of that, Goodall decided to keep to her own and neighboring valleys, in the hope that in a smaller place she would learn the patterns of chimp life more quickly.[26] The decision may have been forced on her by an attack of malaria in mid-August that flattened both Goodalls for a week and left Jane too weak to do much more than struggle up the nearest open slope to rest and watch.

Being forced to curtail her daily wandering proved to be a lucky break. On her first day out of sickbed, Goodall got as far as a rocky peak, which had a superb view of the Kakombe valley slopes. Fifteen minutes later she saw three chimps in a burned area nearby staring at her before calmly moving off. Later a group fed in a fig tree on the opposite slope; they saw her—she was very conspicuous on her outcrop—but again just stared. In just a few days she saw more chimps from her rocky perch than she had in a month of random forest walking. It turned out that the peak—soon honored as "The Peak"—was the best place for watching chimps in all of Gombe. She began spending every day there, dawn to dusk, moving up from her own nesting place, in camp, and waiting motionless and silent for chimps to move down from their nests higher up the slope. Soon she was leaving a blanket and provisions at the Peak so she could spend the night if the chimps were nesting nearby. She was also able to shed her two encumbering minders, and the chimps were less alarmed by one lone human than by three.[27] This way of inhabiting and foraging the chimps'—and now her own—forest home was better adapted to the ecology of data foraging than the random walk.

Goodall's Peak routine also proved a better way of habituating the chimps to her presence than following them in the forest. Being predictably and visibly in one place day in, day out, she became a familiar part of the chimps' faunal landscape, rather than an alien encountered unexpectedly on a forest path. It was surprise that gave animals fright, because it was dangerous predators that hunted by concealment and surprise. Habituating was the opposite of stealth: sitting calmly in plain sight on her Peak, Goodall signaled in the behavioral language of the chimps themselves that she was a peaceful resident and no threat. First with not very good binoculars and later with a high-power telescope, she could see far more of chimp life from afar, especially of shy mothers and offspring, than she could in thick vegetation, though being closer to the animals was emotionally more rewarding.[28] When she saw chimps move out of the fig trees in the valley, she would move down to note the kind of fruits they were eating and perhaps watch from an unthreatening distance chimps at rest and at play in the valley bottom. In this way she became acquainted with the chimps' habitual feeding and resting spots and routes through the bush. Within just a month or two, as she later put it, she knew the maze of ravines and paths as well as London taxi drivers know their city's maze of streets.[29] The chimps still ran off when she got anywhere near. But gradually she perfected the art of getting close: approaching openly but casually, as if the animals weren't there; reading vocal and gestural signs of rising distress when she got too close; calming her subjects by feigning indifference or imitating their behaviors—digging for insects and pretending to eat them, resting, scratching, sleeping, making little noises of chimp contentment. Her "baboon act," Goodall came to call it (chimp act, really).[30] By impersonation and assimilation she became in effect a resident forest animal.

In the two months following Goodall's discovery of the Peak, the chimps became more and more tolerant of her presence. Passing by her on the Peak, they ceased to pay much attention, so long as she didn't move or stare and didn't try to follow. At first they stopped and stared before moving on, then simply ignored her. Encountering her unexpectedly on a forest path, they did not instantly flee, so long as she kept a distance of sixty or eighty yards. That distance gradually diminished, until she could sit quietly near groups and observe the details of their behavior. She soon began to recognize individual animals by physical appearance—which in chimps is distinctive—and then by quirks of personality, beginning with the soon-to-be famous "David Greybeard," whom Goodall first encountered the day she came down with malaria, and who became her guide into the chimp community—Dante's Virgil, William Whyte's Doc.

FIRST DISCOVERIES

Though habituation was of all the practices of observing the one with the greatest long-term potential, it was also the slowest and, for a visitor on a schedule, the most risky. Goodall had funding for just four months and knew full well that renewal would require substantial results. She had in three months accumulated more good data on chimps' daily lives than anyone had previously. But she had made no striking discoveries, and she had only to the end of November to come up with something that would show Leakey that she could do scientific work despite her lack of formal training, and give him what he needed to get her grant renewed.

She did make dramatic discoveries, but perilously close to the end of her four-month trial. These were, famously, that chimps ate meat, and that they deliberately made and used tools. On October 30 she observed a chimp—David Greybeard, again, she later realized—eating meat. Sitting on the Peak waiting for chimps to arrive to feed in the valley below, she saw a great commotion in the trees: excited chimps, angry screams; one holding and eating something pink, and others begging, cajoling, snatching what they could; angry bush pigs milling about on the ground and chasing chimps. The pink was meat, Goodall could see, most likely a baby bush pig. She had not witnessed the actual killing, but it was clear from the excitement and the begging that a taste for meat was no individual oddity but a general trait. Watching the scene from her Peak was for Goodall like reading a detective story and following clues to the denouement of whodunit and how. It was "comical, fascinating, and scientifically valuable," she wrote her family.[31] A chimp village drama.

Goodall's first sight of toolmaking came just a week later, on November 4. After a frustrating morning clambering up and down three valleys, she was heading for the Peak, weary and soaking wet—the fall rains had just begun—when she saw at sixty yards a chimp probing a termite mound with a long grass stem and eating something off it. When after an hour he left, Goodall investigated and found crushed termites and discarded stems; and when she pushed a stem into a hole in the mound and drew it out, she saw it was covered with termites. It was obviously a significant discovery, so before leaving the place she fashioned a crude blind for future watching. In the following weeks the same chimp—yet again David Greybeard—revisited the mound with others, and Goodall watched them break open a hole in the mound, collect the right kind of grass stems, strip off the leaves, and bite the stems to the right length, keeping several on hand as spares. That was the new and really significant observation: chimps did not just use found objects as tools, as many animals do, but fashioned tools with skill and forethought, in the human way.[32]

It was habituation and all-season resident observing together that enabled Goodall to see what others had not, because it increased the likelihood of seeing activities that were not uncommon but sporadic and thus unlikely to be seen by commuting or visiting observers. Termite fishing, it turned out, occurred only in the first months of the rainy season and especially in November, when termites performed their nuptial flight—thus signaling to chimps their availability as food—and in preparation for that flight extended their tunnels close to the surface of their rock-hard nests, where chimps could break through to fish. Had Goodall left Gombe just a few weeks earlier, she would not have seen them making tools. Likewise with carnivory: she later learned that it, too, was episodic: one chimp making a kill would start a brief fashion for meat eating. Goodall was lucky, but made her luck by patient devotion to resident observing.

More serendipitous was the visit to Gombe just three weeks earlier by George and Kay Schaller, who alerted Goodall to the idea (or reminded her) that our chimp cousins might, like us, also be carnivores and toolmakers. (The idea was familiar, though there had been few reliable sightings.) Leakey had sent the Schallers to Gombe, ostensibly to give Goodall the benefit of their experience with habituating gorillas, but probably also to show some trained ethologists a good place for watching chimps, in case Goodall did not deliver. George Schaller intimated that they saw Gombe as a possible research site. He also advised Goodall that seeing chimps eating meat or making tools would justify a whole year of forest walking and watching. It is likely, as Dale Peterson suggests, that Schaller's advice made her "more than usually attentive" to seeing meat eating and toolmaking.[33] But it was her practice of resident observing that put her in the right places at the right moments to see and understand. Just three weeks later she was on the boat to Nairobi.

Leakey, predictably, was ecstatic at Goodall's reports of meat eating and especially of tool using, which he proclaimed would cause scientists to redefine either "tool" or "human"—an exaggeration, but quintessential Leakey. Now armed with facts, he set to work to get Goodall the academic credentials that would give her and her observations the authority of scientist and science. He engineered her acceptance into the PhD program in animal behavior at Cambridge University, despite her lack of the required BA degree, and laid plans to send her to the Yerkes Primate Center in Florida to study captive chimps in semi-natural conditions. He also persuaded the National Geographic Society, which had just agreed to underwrite the work at Olduvai, to keep his "research assistant" (an upgrade from "personal secretary") at Gombe turning up facts that were "fantastically important for science." Cambridge was postponed, and in January 1961 Goodall was back at Gombe, now alone with her new and first-rate Kenyan tracker, Saulo David, at the Peak and on forest trails, walking, waiting, and watching.[34]

The rainy season was a miserable one for fieldwork: wet and cold, with dense grasses twelve feet high. The chimps broke up into smaller foraging groups and were harder to find, and exhausting dawn-to-dusk walking and watching was often fruitless. Yet it was in this season of discomfort that habituation really began to work. The rain seemed to make the chimps less fearful of Goodall's approach, though as they became less afraid they also became more aggressive. There were several frightening close encounters. Goliath, a huge male, threatened from a tree just above her head, then charged, veering off at the last moment. Another male hit her on the head as she lay pretending to sleep. (Her attacker, Rudolf, she learned later was an exceptionally ill-tempered animal.) Even the mild David Greybeard became aggressive when overexcited. Yet the stage of aggressiveness was temporary, and by the time the rains ended, in June, the chimps had come to simply accept her "as part of their normal, everyday landscape. A strange white ape, unusual to be sure, but not, after all, terribly alarming."[35] In a reversal of Georg Simmel's formulation, she was strange but to the chimps no stranger.

By summer Goodall knew about ten of the males individually, though the "women" were more evasive, and the "children" indistinguishable one from another. Fieldwork had become reliably productive and a pleasure, as she reported to her family: "Chimps are <u>extremely</u> friendly—I don't mind how far I scramble after them now, because I know that there is a very good chance that they will <u>not</u> go away when I eventually reach them."[36] She could now sit quietly within ten yards, watching chimps carrying on as if she were not there. A place that had been "alien, strange, confusing—a challenge" was now familiar. "The hills and forests are my home," she wrote. "And what is more, I think my mind works like a chimp's, subconsciously." When she chose a forest track from the maze of tracks, chimps would turn up, even when the track she chose was not the one she thought it was.[37] She learned to adopt the patterns of chimps' daily lives in much the same way that a newcomer to a village would learn to live as villagers do: by observing and making themselves part of the scene. The chimps' home had become hers as well. Her science thus became truly residential.

Goodall's work in the rainy season also brought a third major discovery: the famous "rain dance." On January 31 Goodall was close to a feeding group in the forest, when it began to thunder and pour rain. The chimps moved up the slope into an open area, when suddenly the large males began to charge wildly down the slope, tearing off branches and flailing about, and leaping into trees and hurling themselves down. The rest of the group meanwhile sat in the trees like an audience witnessing a dramatic spectacle (which it certainly was). "I don't think I have ever watched any performance which gave such a thrill," Goodall wrote her family. "Primitive hairy men, huge and black on the skyline, flinging themselves across the ground in their primaeval display of strength and power. . . . Can you begin to imagine how I felt? The only human ever to witness such a display, in all its primitive, fantastic wonder." After thirty minutes, the action stopped and the group wandered quietly off, one male looking back at Goodall as if taking a curtain call. She wondered if the whole performance had been meant for her.[38]

The "rain dance" was for Goodall a more significant discovery than either meat eating or toolmaking. Partly it was the sheer emotional intensity of the drama, heightened by the theater-like setting of the forest glade, which evoked conventional responses to staged dramas; but it was more than that. The rain dance opened up a novel realm of scientific meaning that Goodall's earlier discoveries had not. That primates were carnivores and toolmakers were plausible conjectures, so Goodall's observations confirmed what was expected. Not so the rain dance: it was totally unexpected behavior (Goodall witnessed it only twice more in the next ten years), and it revealed in a most vivid way that chimps—like humans—had a kind of culture, of which scientists had had no inkling. So novel was that side of primate life that the science had as yet no proper words for it. The phenomenon was at first more accountable in the language of dramatic art and fantasy than of science.

In fact, Goodall's language in her letter echoes a passage from one of her favorite books, Edgar Rice Burroughs's *Tarzan of the Apes*, in which the young Tarzan, in "a small natural amphitheater" in the jungle, participates with the wild and hairy ape-men in their "fierce, mad, intoxicating revel of the Dum-Dum," from which, Burroughs posits, all the ceremonials of modern church and state descended. The Dum-Dum was also the occasion when Tarzan, who had been treated by the ape-men as an alien creature, tolerated yet persecuted, was accepted as a full member of the band.[39] I doubt Goodall had this passage consciously in mind when she evoked the rain dance for her family. Yet the resonances between what she described and the literary convention suggest a connection of some kind. Was a story the default way of understanding an incomprehensible and intensely emotional experience—a first step? And did understanding chimp display as theater, a familiar activity of human social life, then open the way to thinking of chimps' group activities generally as a kind of culture?[40]

THE BANANA TREE: PROVISIONING

However rewarding emotionally, forest walking and watching was a science of random encounters, unpredictable and mostly low yield. Goodall learned to recognize the most distinctive chimps as individuals, but groups were too variable and encountered too sporadically to get systematic data on identities and social relationships. That situation changed dramatically with the practice of provisioning: putting out bananas to attract chimps in large numbers to camp, where their social interactions could be observed continuously and up close. This was a data-rich practice—at times almost too rich—and over time it yielded data on virtually every chimp in the Kasakela community and their webs of relationship—the "total situation," to borrow a term from sociology. Although many ethologists condemned the practice as unnatural, provisioning was in fact a range of practices, from the unabashedly artificial to the near natural. Goodall's provisioning evolved by trial and error from the one to the other.

It was the chimps themselves who initiated provisioning, in late February 1961, when David Greybeard appeared unexpectedly in Lake Camp to feed on the nuts of a palm tree that had just become ripe. In the next weeks and months, David and then other males came regularly to camp, occasionally in pairs or small groups. As the chimps got used to humans, they included the camp's fruiting trees as a regular stop in their foraging itinerary—a kind of self-habituating. For Goodall it was a treat to watch chimps comfortably from a tent instead of in the wet forest. However, there was just so much to be learned from watching one chimp after another stuff itself, so at first she took their visits more as entertainment for her staff than as science: a temporary curiosity that would end when the palm nuts were gone. She would sometimes linger in camp just for the pleasure of seeing her friend David up close, but she did not at first make Lake Camp a regular part of her own data-foraging rounds.

That changed, however, when one day David walked nervously to within five feet of her, then dashed up and snatched a banana from the table. Goodall then instructed Dominic to set out bananas to keep the chimps coming when the trees were no longer in fruit, and the chimps quickly acquired a taste for the exotic fruit, returning often and then every day. Goodall in turn added visits to camp in her round of forest walking, first occasionally, then every day. Visits to camp thus became a regular habit of foraging for observer and observed alike: chimps could count on finding good fruit, and Goodall was assured of good data. It was a gradual shift in field practice from one mode of residence and observing to another.[41]

Banana provisioning greatly accelerated the process of habituation, as chimps learned from one another to tolerate close contact with humankind. David Greybeard was crucial, bringing first his friend Goliath, the community's dominant male, and Goliath's timid sidekick, William; then others as well. In camp David had a calming effect on nervous or frightened individuals, as he did in chimps' forest gatherings—he was the community's peacemaker, who got along with everyone. Goliath was the second to become "tame," in 1962, though Goodall never trusted him. He was "slightly mad and not very intelligent," so was unpredictably violent.[42] David, meanwhile, became very familiar indeed. One morning Goodall awoke to find him sitting by her bed eating a banana he had pilfered from the storage box, and later he took a banana from her hand for the first time—a thrilling experience. In 1963 females with their offspring also became regular participants in the camp banana feasts, beginning with the famous Flo. Flo was old, worn, and strikingly ugly, yet an exceptionally able and successful mother, and a sexual magnet for all the Kasakela males. During her long periods in estrus, or "pink," she brought with her into camp crowds of male suitors, who then kept coming for the bananas after Flo was no longer in pink.[43] Thus through the normal activities and behaviors of group life, chimps one after another became habituated to a place of human residence and made it their own.

Goodall and others have remarked on the ecological nature of provisioning. As Dale Peterson put it, "The banana-feeding regimen was never substantially different, ecologically, from a prolific, unusually reliable fruit tree in the forest." Chimps reacted to the banana tree as they did in the forest to fruiting fig or msulula trees, congregating to gorge until the fruit was gone. The difference was that the new tree of Lake Camp was always in season and never without fruit enough for all comers. And when the feeding operation was moved to a new camp farther up the ridge, the chimps were right behind. "So far as they were concerned, the bananas, after an exceptionally long fruiting in one area, had become ripe in their strange underground boxes somewhere else."[44] Provisioning thus seemed a natural amplification of normal behavior, not an intrusive scientific artifice.

Camp life soon came to revolve around its visitors. Bushes and trees in the provisioning site were cleared to improve photography and filming. (Funding from the National Geographic Society came with demands for regular, high-quality product.) Staff vanished at the low whistle that signaled the arrival of chimps; cooking and eating stopped; and everyone stayed in their tents to observe, record, and film.[45] The chimps were in effect reclaiming Lake Camp, and then Ridge Camp, as part of their own landscape of residence and foraging. Camp became a place of coresidence, as the forest had. Observers were inside the object of their study not just in the chimps' forest home, now, but in their own home as well.

Provisioning dramatically changed the pace and scope of the scientific work, as chimps were drawn to camp from farther afield and in larger and more varied groups. Around the banana tree, behaviors could be seen repeatedly that in the forest were seen rarely and serendipitously. A range of greeting behaviors was systematically observed in their particular contexts of use. Individual variations in communicative gestures and vocalizing could be observed and heard up close. Interactions between members of ever-changing groups produced systematic evidence of individuals' personalities and social relations. Competition for access to food revealed how social rank order was maintained and altered through encounters between individuals of different status. In the forest, where individuals and groups were encountered haphazardly, observers were never sure whether observed behaviors were general habits or just idiosyncrasies of the individuals they chanced to meet on forest paths. Generalizing was more secure with data gathered at the banana tree. It also became possible for the first time to get data that could be analyzed numerically and statistically, like the data obtained in a laboratory or colony of captive animals. As Goodall wrote to the botanist Bernard Verdcourt in 1963, "Honestly, Bernard, I am learning so much in such a short time that I sometimes wonder how much longer it can go on. I mean—there must be a limit to the amount I can go on learning about chimps—such large amounts of new things, anyway. Of course, one would expect to go on learning in dribs and drabs for years and for ever."[46]

By 1964 the volume of data produced was straining Goodall and her staff to the breaking point. Recording every little thing every chimp did in camp every day, while keeping up the forest walking and watching, was almost impossible. As Goodall wrote Melvin Payne, her contact in the National Geographic Society, "[T]he overall work has increased so tremendously that we are all three [Goodall, her husband, Hugo van Lawick, and assistant Edna Koning] going flat out from about 6.30 a.m. until nearly midnight each day simply in order to keep everything up to date. . . . It is wonderful to be acquiring so much information—but horrifying when one thinks of analyzing it."[47] Goodall herself was spending more and more time in her tent dealing with records management, and less and less in the forest, as assistants were delegated to do what had been Goodall's sole work and pleasure. "I can hardly bear to do this," she wrote her family in August 1964. "[M]emories of the valley are so vivid, and these days it is all work work work with little time to roam through my mountains. I wouldn't wish it otherwise though. Being able to observe the sort of behaviour that we see these days, and beginning to understand so much about the social pattern, is worth everything."[48]

As provisioning was to the chimps a superabundant ever-bearing banana tree, so was it for their human observers a *data* tree, bearing scientific fruits abundantly and without fail in all seasons. The practice had an autocatalytic quality: the more fruits were consumed, the more were supplied; the more data gathered, the more there were to gather. The more Goodall and her team learned of individuals' personalities and relationships, the more there was to learn about situational variations in behavior and the detailed structure of chimp society. This autocatalytic property is less typical of field science than of science in labs, where new kinds of instruments and experiments are devised, and every measurement or experiment that is done generates more to do. (Mapping genetics with standard organisms, for example: measuring map distances between mutant genes produced more mutants to be mapped.[49]) So it was for resident observers at the data tree: the more they knew of chimp social structure and its operating principles the more there was to learn.

In one respect, however, provisioning was not unambiguously an improvement on forest walking. As more chimps showed up, it became increasingly clear that they behaved differently in camp than in the forest. Congregations were larger and more disorderly. When high-ranking individuals did not have first pick of desirable fruit, as they did in the smaller forest groups, normal mechanisms of deference and order failed. The chimps got overexcited and aggressive, and fighting became chronic. Goodall estimated that about two-thirds of observed instances of aggression in camp were caused by the abnormal situation of provisioning.[50] Normal patterns of forest wandering also changed, as chimps began to nest and stay near camp, causing more excitement and aggression.

The human life of camp was also disturbed. As the chimps lost their fear, they became more assertive with humans and as casually destructive in camp as they were in the forest. They began to raid tents in search of bananas and tasty things to chew: cardboard, clothing (for the salt), shoes, knapsacks, tent canvas, chair legs—they would gnaw just about anything. They broke open banana boxes and anything else they thought might contain bananas, and snatched and ate eggs (a favorite delicacy) and chickens. Tents had to be closed and everything put in secure containers when the chimps arrived. More alarmingly, the chimps began to raid the huts of the African fishermen near Lake Camp, and when Leakey got wind of that, he raised the alarm. Africans would not tolerate chimps' depredations as Goodall did, and if they or the chimps were harmed or killed that would put an end to the Gombe project. Goodall assured Leakey—somewhat overoptimistically—that the chimps would stop coming to camp if the bananas stopped for a few days: they would "presume the tree was out of season and move off. They will not look for them anywhere else. Why should they?"[51] Because they were chimps, was why; and they did not move off.

A move from Lake Camp, in 1964, was a partial remedy. The new Ridge Camp was half a mile up the valley from shore, so chimps no longer invaded fishermen's huts. But the more natural forest site did not solve the problems of crowding and aggression. Being more at ease in the forest, chimps became even more aggressive in their quest for bananas, shoving Goodall and her staff aside to dive into the banana box, and smashing equipment hoping to find fruit inside.[52] Efforts were made to thwart hogging and hoarding by hiding bananas around the site, but that only encouraged chimps to search even more imaginatively and destructively in tents. Conflict between observers and observed was inevitable in a closely shared space: its human residents wanted an optimal situation for living and work; chimps wanted all the bananas they could eat, on demand.

Controlling the banana tree became a seesaw war of wits: the dominant primates would devise clever new controls, and their clever cousins would thwart their every move.[53] Goodall and her staff turned first to technological solutions—the first recourse of humankind. They presented bananas not scattered on the ground but placed in half-buried concrete and later steel boxes equipped with mechanisms that could be opened and closed from a distance. There were enough boxes initially (thirty-eight) to feed all comers, and boxes were widely spaced to discourage fighting and give every animal a fair share. But chimps are observant and handy (like us), and quickly learned to operate or break the release mechanisms. Flo's clever offspring, Fifi and Figan, led the way, and others followed their lead. They learned to dig out and break buried cables, which the camp staff then buried in concrete. However, chimps could still see boxes being stocked, which stirred frenzies of anticipation. So a system was devised of boxes that were opened electronically from tents and stocked with fruit from a trench in which observers could move unseen. That thwarted everyone except David Greybeard, who would take a nearby person by the hand, pull him or her to a box that smelled of bananas, and keep a firm grip, getting more and more excited, until the box was opened. When David arrived in camp, everyone had to stay indoors until he left: "Talk about being in a cage!!!!!" Goodall expostulated.[54] It was resident science of a perversely unanticipated sort.

The ultimate solution to the problems of provisioning was ecological: to make the banana tree more like an actual forest fig, fruiting unpredictably and briefly. Chimps were given bananas "in a way most similar to a natural food supply, and so as to affect as little as possible [their] . . . social behavior."[55] That end was achieved by slow trial and error between 1962 and 1968. The schedule of feeding was varied to discourage chimps from hanging around. Feeding days were alternated with nonfeeding days, but that pattern was predictable. Feeding days were then selected randomly, in the hope that chimps would visit randomly, but on feeding days the frenzy and fighting continued. Finally, the number of boxes was reduced (to twelve or eight), and boxes were baited and opened on a schedule tailor-made for each visiting group, so that any chimp was fed only once in ten days, and only if it had been in camp less than fifteen minutes and in a group of less than six. With that combination of machine and ecologic the human residents of Ridge Camp finally got the upper hand. Fewer chimps arrived each day (ten to twenty)—"a convenient number for purposes of observation"—and in smaller groups; and chimps who nested far from camp ceased to come.[56] The chimps thus reverted to their normal foraging behaviors, as the banana tree became just another tree to check on their wandering forest rounds.

Naturalizing the banana tree also rebalanced the human ecology of data foraging. Stemming the deluge of data relieved overstressed recorders with little cost in loss of data. In fact, an abundance of data was no longer as vital as it was in the early days, when facts were sparse and more was always better. By the later 1960s Goodall had a backlog of data awaiting compiling and analysis (a task she dreaded), so it made practical sense to gather selectively to address particular questions and hypotheses—especially after the future of science at Gombe was secured by the establishment of a permanent field station just uphill from Ridge Camp. As unlimited but unnatural provisioning had suited an economy of scarcity and insecurity, so did an ecologically naturalized practice suit one of security and abundance. Integrating the banana tree into the normal ecology of the forest also removed any lingering taint from the data it produced. As Dale Peterson put it, there was nothing "egregiously unorthodox or blatantly unscientific" about Goodall's practice of provisioning.[57] At Gombe the ecologies of fruit and data foraging were made intricately congruent.

GOMBE RESIDENT SCIENCE

Naturalizing the data tree also restored a balance between provisioning in camp and forest walking and watching. As the demands of crowd control and data management subsided, observers were once more free to follow chimps on their meandering forest ways—all except for Goodall herself. Officially credentialed (in 1965) as Dr. Goodall, she inherited the status of principle investigator from Leakey, along with full responsibility for the finance, administration, and politics of the Gombe operation—a job that kept her in camp all day every day. Her young coworkers were eager to walk and watch, however, and Goodall was glad to afford them the experience that she had found so rewarding. In fact, the years of all-out provisioning had made forest walking more productive and expansive than it had been before.

With the comprehensive knowledge of her subjects gained by provisioning, Goodall could correct errors she had made earlier by generalizing from the sporadic data of forest watching. For example, she had initially concluded that the bipedal swagger response was typical of chimps, but it turned out that just two or three individuals were responsible for most of what she had seen.[58] Long-term observing in both camp and forest also made it possible to follow individuals' life histories over time, in some cases from birth to maturity, in others from maturity to death. (The death of worn-out Goliath was sad; the death of David Greybeard, personally devastating.) As chimps had distinct personalities, so did they have distinctive lives shaped by particular family and social relationships, and by random occurrences. Chimps had stories and biographies that could be told much as we tell stories of our own lives.[59]

The combination of systematic data gathering in camp and forest watching also revealed a surprising range of individual variation in chimp social repertoires. The language of chimp communication—in demeanor, posture, gesture, touch (touch was especially important), and vocalizing—had definite conventions. And these were altered and adapted to particular social situations, very much as the conventions of human interaction are.[60] For example, four adult males could display variously appropriate responses to the same situation: chimp A might stare with his hair bristling, B look away and shake branches, C perform a bipedal swagger, and D sit with shoulders raised and arms held away from his body; and these responses could vary even more if females were present.

The most significant change in this renewed forest walking was its larger scale. No longer restricted to what one observer (Goodall, usually alone) could cover in a day, Goodall's acolytes could walk the entire territory of the Kasakela community, not just the four nearest valleys. And because all the members of the community were by then known (thanks to the banana tree), forest walkers could interpret the behaviors of every group they encountered, even those from distant valleys. Observers could work out patterns of group territoriality, and witness the sometimes violent encounters between Kasakela chimps and those residing in neighboring territories. The most dramatic new discovery was the deliberate killing and eating of infant chimps, usually in attacks by males on strange females, but in a singular case by a cannibal mother and daughter who had developed a taste for it. Equally dramatic was the systematic stalking and brutal killing by Kasakela males of a group of males and females who had split off from the community and staked out a piece of the communal territory as their own. These discoveries of chimp cannibalism and warfare were almost too shocking to believe at first, and uncomfortably reminiscent of human nature.[61] They were the dark side of the carnivory and toolmaking that Goodall had discovered in her first months of forest walking. And they were revealed only by the systematic walking and watching of many observers over the entire range of the Gombe forest.

The most important discovery scientifically of this renewed and wider forest walking was that the Kasakela community as a whole had a structure and, so to speak, a sociology. That was a surprise. In the forest and in camp, chimp society seemed to consist of ever-changing groups of two to six individuals. Only family groups of mothers and offspring were stable; others varied widely in their mix of sex and age classes. This social fluidity was one of the first things that beginning forest walkers noticed, and Goodall had concluded early on that the Kasakela community as a whole was as unstructured as the groups. However, that conception turned out to be an artifact of incomplete data. The complete data from provisioning and wider walking revealed the very different picture of a whole community in which every chimp had a definite position and relation to every other one. Every chimp knew its place in this order: not through direct experience of the whole—the whole community was never together at one place and time (as vervets and baboons always were)—but by the cumulative experience of the fluid groups in which they lived from day to day. A chimp that was dominant in one group would not be so in others, and in any group it would know (or quickly learn) where it stood in relation to the others present. In this piecemeal way individuals came to know all the members of the community as well as their own place in the whole.

Goodall and her staff in fact worked out the structure of the whole community in the same way that the chimps did themselves: through the cumulative experience with many small groups. By a combination of provisioning and expanded forest walking, they came to know every member of the Kasakela community, as well as the relationships of each to each as displayed in their behavior in foraging groups. "A chimp community is an extremely complex social organization," Goodall wrote. "Only when a large number of individuals began to visit the feeding area and I could make regular observations on their interactions one with another did I begin to appreciate just how complex it is."[62] The varied gestures of greeting, reassurance, assertion, and deference that groups of chimps displayed upon meeting at fruiting trees or on forest paths were signals that they remembered their past interactions with each other and knew their current position in every possible social situation. These clues to the structure of the whole were there all along but became readable to the observers only when they had witnessed groups and encounters in all their permutations and combinations. Then the whole picture emerged, as the picture does in a puzzle of many little pieces.[63]

Exhaustive observing of groups in camp and forest also revealed the behavioral mechanisms that shaped and sustained the chimps' social order.[64] Connections of family and friendship proved to be as decisive as individual qualities. Dominance, for example, derived not simply from physical strength and aggressiveness, as might appear from dramatic encounters, but also from skill in raising capable young and in cultivating social alliances. Ties of kinship among chimps are strong and lasting, especially between mothers and offspring, and between adult siblings. That became apparent only when Goodall and her staff were able to observe family groups through a full generational cycle. Males with supportive brothers had an advantage in contests for social standing and precedence, as did females with supportive male and female offspring. Flo's dominance, for example, derived not just from her exceptional intelligence and personal qualities, but also from the strong ties she created as an exceptionally effective mother.

Among unrelated animals, standing depended largely on individuals' ability to form alliances of friendship, like the one between David Greybeard and Goliath, who were constant (if unlikely) allies. Chimps spent much time and effort cultivating alliances, using a varied repertoire of gestures of greeting, reassurance, and support—the tool kit of chimp social politics. The social meaning of these behaviors was gradually revealed to observers, as alliances of family and friendship were deployed in contests for food and social status in ever-changing groups. As resident observing revealed a structured chimp society, so too did it reveal the behaviors that maintained and slowly changed it. The naturalized banana tree and expanded forest walking together constituted the material and social tool kit of a chimp sociology.

Resident observing for Goodall was not just a visual tool but also an emotional or affective one. It is common for scientists who study animals to develop strong feelings about their subjects, of course. Uncommonly, Goodall made those feelings serve the purposes of empirical science. In a scientific culture that took affect and reason to be dichotomous categories—good science is dispassionate—Goodall pursued a science in which personal feelings deepened empirical observing and understanding: not feeling *and* science, but feeling *as* science. The balance was tricky to maintain. Dale Peterson writes that though Goodall "would portray herself as ambitious at science," it was her passion for animals, and for her chimps in particular, that sustained her work at Gombe.[65] True. The point, though, is that passion was not separate from the science, but part of it. Goodall's books and letters reveal how affect worked as science.

The intimacy of coresidence was the foundation of affective science. It was one thing to recognize chimps as individuals, and another to *really* know them, as we know close friends and neighbors. Intimate moments in the forest sitting near groups of chimps were important, but it was the sustained cohabitation in camp in the period of provisioning that made real intimacy possible. For example, Goodall had known Flo by sight for a year from forest watching, but in a year at Ridge Camp Flo and her family became, in Goodall's own words, "an integral part of our lives. We learned a great deal about their behavior by means of objective recording of fact, but we also became increasingly aware of them as individual beings. Intuitively we 'knew' things about them which as yet we could not begin to define in scientific terms. We began, though indeed 'through a glass darkly,' to understand what a chimpanzee really is."[66] The intimacy and affect of coresidence in "Chimpland"—as Goodall headed her letters to family and friends—thus became empirical science.

Famously, Goodall made emotional contact with individual chimps by interacting with them in their own language of touch and gesture, as if she were experiencing the emotions elicited in interactions between the chimps themselves. That is how Goodall's "baboon act" worked: by imitating the gestures and demeanors that chimps used to calm and reassure, she learned the language of chimp affect. It was playacting, but it became a kind of virtual or imagined participation. A transformative moment for Goodall was when David Greybeard gently took a banana from her hand—a hint of communication between species.[67] Even more stirring was the occasion when David Greybeard briefly held her hand. She had been following David through the forest when he seemed to vanish but then reappeared, as if waiting for her to catch up. They sat side by side "eating leaves." Goodall then picked up a ripe palm nut and held it out to David, who gave it a scornful glance and turned away. So she held it closer, whereupon David suddenly reached out and held her hand with a firm warm pressure for about ten seconds, then withdrew, glancing disdainfully at the nut and dropping it on the ground. Goodall, who by then could parse chimp gestural language, grasped what David's gestures meant. Chimps hold hands when a subordinate wants reassurance and a dominant wants to reassure; David didn't want Goodall's gift but took her outstretched hand to be a sign of wanting reassurance, which he gave.[68]

Goodall remembered this drama as "the most significant event in my life": the moment when she understood both intellectually and emotionally just how narrow the species gap was between chimps and humans.[69] She felt on her own body the psychosocial effect of chimps' language of social interaction. Goodall had long known the gestural vocabulary and syntax, but it was in personal contacts and virtual participation in acts of communication that she *really* knew how chimp relations worked emotionally. It was an experiential science more vivid and rich than simple observing, yet no less empirical. That science could even be used in quasi-experimental ways, as when David suffered a cut in a fight, together with Goliath, with a pack of baboons. Peering at David's cut, Goodall put out her hand to groom his back, whereupon Goliath came menacingly at her with his hair standing up, as he did whenever his friend and ally seemed threatened. Goodall withdrew her hand, and Goliath returned to his seat. The experiment was repeated, with the same result. "Most spectacular and very touching," Goodall thought.[70] It was also most revealing of the operating principles of chimp society. Such episodes call to mind William Whyte's experience of bowling and of feeling in his own nerves and muscles how emotions shape group structure and behavior; and how in his experiments with Long John and Doc, he turned that affective knowledge into predictive science.

Goodall was likewise deliberate and disciplined in using affect to scientific ends. Empathy and intuition were tremendously valuable in understanding complex social behaviors, she wrote, and entirely scientific so long as behaviors were recorded precisely and objectively as they were observed. It was nonetheless an attitude of which many ethologists disapproved. The zoologist Adriaan Kortlandt, for example, disparaged Goodall's field practice as the "Saint Francis of Assisi" approach.[71] For him, and for many others, it was detachment that guaranteed objectivity, not feeling and personal engagement. Goodall was undeterred by such criticism, however, and pointed to our close evolutionary relationship to chimps as reason to believe that insights gained by emotional engagement could be trusted as evidence. Primatologists should not shrink from using a perceptual faculty given to them by nature.[72] As the ethologist Fraser Darling had written some years earlier, "In some instances I feel that the most simple explanation of an act of behaviour is to follow the bare outline of our own mental processes in such a situation. Who are the people with whom the higher animals are most serene, and who achieve most success in their management and training? Not those who look upon them as automata, but those who treat them as likeable children of our own kind."[73]

Goodall's disciplined use of affect was not the anthropomorphism of those who identified so closely with their animals that they sought to be one of them, like Dian Fossey, who wanted to *be* a gorilla. Her gorillas, she believed, understood her better and were better friends than any of the humans in her troubled life. (To one onlooker—a spirit medium, as it happened—she appeared as a gorilla in clothes.[74]) Goodall never wanted to *be* a chimp, though she did try out chimp tree nests (comfy) and chimp foods (nasty). She felt "a closeness, an awareness, an empathy, a respect for, love of" her chimps, yet was always the critical empirical observer. She did for a time use the human category of "friendship" (*My Friends the Wild Chimpanzees*, she titled her first book), but with experience grew more circumspect. She observed that chimps, seeing her after a long separation, displayed none of the affection that they did with each other, and signaled recognition only by ignoring her. Judged to be alien but harmless, she was tolerated. Only David Greybeard was a friend in a deeper sense, she thought, having reached out to her with expressive gestures.[75] On reflection, however, she concluded that "friend" did not apply even in his case, because friendship is a reciprocal relation, and her relation with chimps was one-way. She depended on them in her scientific and emotional life, whereas they depended on her for nothing. As Sy Montgomery put it, "Jane is not one with chimps. She is a visitor in their world but 'not a citizen.'"[76] Her "anthropomorphism" was an affective practice of empirical science.

SCIENCE AND LIFE EXPERIENCE

So how did Goodall do it? From what sources did her distinctive practices of habituating and affective connecting derive? Not from anything in her scanty formal education, obviously; nor from Leakey, who never visited Gombe and offered no advice; nor from ethologists, with whom Goodall had little contact in her first years. Nor did she discover the practice in the field: she arrived there with it. We see it before she got to Gombe, in her observing of Lolui Island vervets. Its most likely source is Goodall's previous life experience. Dale Peterson and others have pointed to Goodall's early fantasies, inspired by her reading, of living and communing with animals: "Intuitively, she moved to achieve an intimate connection with her chimps: to participate in their lives and perhaps thereby to follow her childhood dream of living close to nature and wild animals in the style of Doctor Dolittle. Following such dreams and intuitions fortunately happened to converge with reasonable science."[77] These literary fantasies were certainly a motivating force for Goodall to get to Africa; but once she was there, what specifically in them could translate into working practices of habituation and inside observing? Doctor Dolittle spoke animal language effortlessly, and Tarzan was raised as an ape: no models here for slow and arduous habituation.

Peterson also points to young Jane's outdoor life in Devon. In Africa, he writes, she "intuitively" chose "the sort of thing she had done as a young girl observing birds and small wild animals along the Bournemouth cliffs, proceeding openly rather than secretively and taking care not to alarm her quarry. She thus had only to learn wild monkeys' tolerance for such an approach: how close to go, what movements to make or not, what clothes to wear."[78] Yet the possibilities of engaging the small animals of the English countryside were limited, and what Jane actually did in those coastal ravines isn't clear. It's all a bit vague for cause and effect. And there were activities closer to home that connect more directly to her later science: namely, her intense interactions with companion species—domesticated animals—and with the people in her life.

Goodall, it is clear, had a knack for relating to domesticated animals, especially horses and her adopted dog, Rusty, who belonged to a local hotel but adopted Jane and spent every day with her. Goodall later recalled how much she learned of animal personality and behavior from Rusty. She learned to read and use the body language of gesture and demeanor by which animals manage relations among themselves and, in the case of companion species, with humans. Rusty, she later recalled, was an exceptionally intelligent and responsive dog: he learned tricks without being taught (it seemed), and was the only dog she ever knew with a sense of justice, becoming visibly aggrieved when wrongly chastised.[79] Young Jane, it seems, was quick to learn the gestural language of animal-human communication, reading Rusty's emotional states and by subtle cues eliciting behaviors—"tricks"—that pleased.

With horses Goodall also displayed an exceptional rapport, from her early teen years, when she did stable work in exchange for riding lessons and spent hours afterward just hanging out in the pastures with the horses. In Kenya she handled animals that even the horsey set regarded as difficult or dangerous. At a pony meet with the Leakeys while being vetted for the Olduvai expedition, Mary Leakey suggested Goodall ride her own pony, Shandy, without mentioning his bad habit of walking backward and making his rider look a fool—it was clearly a test. But mounting the pony Goodall sensed something odd in its behavior and, dismounting, discovered that he had a painful saddle sore. Leakey was impressed, and Goodall passed her test. On another occasion, at a hunt, Goodall rode with notable success a huge horse with a reputation of being too dangerous to ride.[80] Goodall's ability to read and manage volatile and unpredictable animals translated directly into her work and life with chimps.

Though chimps are far from being a companion species, they are a cousin species, and mingling with them required the same abilities that Goodall had mastered with dogs and horses: reading the signs of their emotional states and by her own gestures and demeanor returning signs that the animals would understand. Peterson writes that Goodall's exceptional rapport derived from an intuitive ability to imagine and respect animals as individual creatures with feelings and personality.[81] I agree, so long as intuition and empathy are seen not as innate qualities but rather as social practices learned through everyday experience. We can see in her ability to read and manage domesticated animals what made Goodall such an effective observer of free-living chimps.

I would extend this argument to Goodall's relations with the people in her life. She was, it is clear, unusually adept in human relations. She loved the situations of family and social life and the reading, managing, and creating of them: "attuned to the drama of everyday life," Dale Peterson wrote, "socially outgoing yet very self-possessed." Her family circle of forceful, independent, and supportive older women—her grandmother Danny, mother Vanne, aunts Olly and Audrey—was an exceptionally rich social and emotional milieu. Jane had a quick and sure grasp of social situations and was often the one to initiate family activities—traits indicative of leadership in small groups (recall Whyte's Doc).[82] Her friend Clo remembered her ability to make situations turn her way: "Occasionally if she wants something she looks at you in a particular way, and she gets calmer and deeper and deeper, and you sort of give in. . . . Oh yes, she uses her gaze—I call it her 'look'—a lot, with great success."[83]

Goodall was distinctly dramatic and drawn to dramatic personalities and situations. At the Coryndon Museum, Vanne found the whirlwind of drama around Louis Leakey exhausting. "I have never been embroiled in so many intrigues," she wrote her sisters, "so many cross currents of fevered emotions, so many secrets to be kept from almost everyone, and so many comings and goings which may or may not be divulged. One's head literally reels with it all." Jane, the mistress of situations, reveled in such scenes. The atmosphere of the Coryndon Museum was "super," she wrote her family, and the staff "charming & great fun, everyone mad." It took all her skill to manage Leakey's recklessly indiscreet infatuation with her, which was embarrassing to her and potentially fatal to her Gombe project. She eased him into the mutually agreeable and safe relation of foster father and daughter.[84]

Goodall also deftly managed relations with her needed but sometimes imperious patrons at the National Geographic Society. A revealing episode occurred during the visit to Gombe in 1965 by the men charged with deciding whether to make Goodall's camp a permanent research station. They were three very distinguished and very alpha males: Leonard Carmichael, experimental psychologist and former secretary of the Smithsonian; Melvin Payne, the society's vice president and Goodall's chief intermediary; and T. Dale Stewart, zoologist and director of the Smithsonian's Museum of Natural History. With no scientific credentials, Goodall was distinctly not an equal among equals, and her visitors made that clear.

Carmichael was on Goodall's case from the start, pouncing on things she said that seemed to him less than properly scientific, like giving chimps names and personalities, and telling Chimpland stories, until, finally, she expostulated that she wasn't speaking a scientific thesis but just conversing. From then on Carmichael talked about the chimps as Goodall did, without worrying about his scientific probity. Stewart, too, was critical of Goodall's field practices and went on about his own pet theory of how chimp and human species parted ways. But he softened when Goodall challenged the science, and the next morning mentioned casually that he was having second thoughts. The two scientists respected people who knew the facts and had the confidence to disagree effectively, and it seems Goodall sensed that quality.

Melvin Payne, the lay manager, was a harder case. He bristled when the idea was broached of an independent Gombe film, and grew livid when Goodall told him that a National Geographic film made for popular consumption should not have been shown to a scientific audience. Goodall fumed but held her tongue and awaited the right moment, which arrived later that evening around the campfire when Carmichael remarked how well the showing of the film had been received, and Payne remarked that should set Goodall's mind at rest. It was "[m]y opportunity, and all unasked for," she recalled. "So I regaled Carmichael and Stewart with the examples of unscientific things which had crept into the film. Stewart was on my side in a flash. Carmichael looked into the fire for a long time, and could only say that anyway it served to interest people in the chimps."[85] Mobilizing scientific allies in an adroitly managed scene, Goodall established her standing as scientist and as woman in charge of the Gombe operation.

Is it any wonder, then, that Goodall the mistress of situations took so naturally to a kind of science that required adept reading and managing of primate situations? Chimps are, like us, a dramatic, emotional, noisy, squabbling species—a roiling troop of chimps is for all the world like a troop of excited schoolchildren.[86] They provoke situations to get food, assert social standing, and face down rivals, just as we do. And it is in just such dramatic situations that observers learn most about the personal and social traits of the creatures they study, chimps or humans. Recall Malinowski's use of village dramas to get his subjects' unedited views, and Doc provoking arguments to elicit information from his group for Whyte. And what was provisioning, but a way of encouraging such situations in a place where behavior could be closely yet unobtrusively observed? Goodall's experience and love of dramas in her own life enabled her to understand and even arrange dramatic events in Chimpland.

There is much evidence that Goodall experienced chimp social behavior in terms of human drama—recall the rain-dance theatrical. Her scientific achievement, Dale Peterson wrote, was "the idea of chimpanzees as actors in their own drama." Her accounts of chimp life are full of theatrical scenes. "The other day I saw 4 chimps being baited by a pack of baboons," she wrote in one. "It was like a play being enacted, in a little open clearing down below. The 4 baboons got closer & closer until the chimps could stand it no longer. The 2 females climbed trees, the 2 males chased after the nearest tormentor." Her husband, Hugo, Goodall recalled, would often say that "it was like being spectators of life in some village."[87] It brings to mind (once more) the ecologist Charles Elton's aphorism that "[w]hen an ecologist says 'there goes a badger' he should include in his thoughts some definite idea of the animal's place in the community to which it belongs, just as if he had said 'there goes the vicar.'"[88] Resident observing of chimps or other animals required an ability to think, feel, and act across species lines.

Much of Goodall's correspondence with her family consisted of swapping stories of human and chimp village life. In the well-known letter describing David taking a banana from her hand, for example, there is a less familiar yet telling passage thanking her family for their letters about their daily life at home: "Super letters. . . . Danny, how you can say your letters are dull and uninteresting I just don't know. Far from it—I was breathless with suspense when reading of Kip's wasp, and horrified when you couldn't find him. . . . Oh no—your letters are very very far from being dull."[89] These reciprocating stories of village life in Bournemouth and Chimpland suggest just how close were Goodall's life and science.

It is striking how much of Goodall's writing on chimps is in narrative form—anecdotes, stories, life histories of individual chimps and their dramatic social relationships. This is especially, and unsurprisingly, so in her popular writing and letters home: these were meant to be dramatic entertainments with herself as the central character. *My Friends the Wild Chimpanzees*, which was written for readers of *National Geographic Magazine*, is an adventure story of a courageous and determined heroine who entered into chimps' world and brought back the stories and the science. Adventure and drama were what her patrons required.[90] More tellingly, stories of sitcom life in Chimpland are also featured in Goodall's scientific works, not just to enliven the science but as the vehicle for organizing empirical evidence and explaining what it meant. Goodall's monographic summation of the Gombe project, *The Chimpanzees of Gombe*, includes, along with tables of data and pie charts of the frequencies of generic behaviors, a full dramatis personae of the Kasakela chimps with names and snapshots, along with all the stories of David, Goliath and William, Flo and her family, and the other characters and chimp dramas that Goodall and her fellow watchers had observed. Analytic and narrative forms of presentation derived in part from two distinct modes of fact gathering: numerical data sampling; and resident observing of particular animals and occurrences.[91] However, it is the sketches of individual chimps' personalities and life dramas that carry the weight of scientific depiction and inference—the inside stories of resident primatology.

This intertwining of storytelling and science is especially striking and effective in Goodall's best-known and arguably best book, *In the Shadow of Man*. This book has been read as another, if superior, adventure story. Peterson, for example, called it "first of all an adventure story, modestly told of courage and determination," and "a compelling piece of popularized science." But I think it is more accurate to say that the adventure and the stories *were* the science. The book was not popularized science but science, full stop. It is a seamless blend of autobiography, empirical observation, and interpretation. Chapters on her coming to Gombe and learning to habituate and observe, and how her work gradually progressed, alternate with chapters on the chimps themselves and the major topics and discoveries as they unfolded: chimp personalities and friendships, the social hierarchy of the community, Flo's remarkable sex and family life, individual development and life cycle.[92] For Goodall, *In the Shadow of Man* was "the real book." As she wrote to Julian Huxley, "[T]hough written for the general public it says I think a good deal more than my scientific monograph. I can't help feeling it is a more meaningful statement."[93] Her point, I take it, was that an account just of the factual and analytic results of her science was incomplete without a personal account of the process and experience of her work and life as a resident observer in Gombe. The residing was an essential and integral part of the science. It was the experience of engaged close observing—the experience of coresidence—that revealed scientifically what chimps *really* are. And *In the Shadow of Man* is an account of coresidence. The same can be said about the science and stories of Malinowski's *Argonauts of the Western Pacific* and *Crime and Custom in Savage Society*, Anderson's *Hobo*, and Whyte's *Street Corner Society*—and, perhaps, of all resident science.

IN THE VILLAGE OF PRIMATOLOGISTS

There was an additional dimension to resident observing in Goodall's case that has not come up so far in this book. Goodall had to learn to reside not just among the forest chimps of Gombe, but also among the zoologists and ethologists of Cambridge University and the academic jungle. For most ethologists—for most scientists—this was not an issue: they learned the codes and manners of professional life through a long and graded formal education before going afield. For Goodall it was the other way around: the Gombe forest was more school to her at first than the groves of academe. Academic science was forbidding terrain for a woman who came out of nowhere with a conspicuous lack of scientific credentials, yet who quickly acquired a factual knowledge of chimps so fresh and rich as to be almost unbelievable. Goodall had, in effect, to habituate ethologists to her unorthodox person and practices, and to habituate herself to their unfamiliar, even alien, knowledge culture. She had to develop an academic equivalent to her "baboon act," using academic language and demeanor to her own ends. She had to seem to munch bitter scientific leaves and fruits without really eating them. She had to learn to participate in a community of ethologists without offending communal mores yet without going totally native. However, her skill and experience in reading and managing situations served her as well in academia as in the ravines of Gombe.

Presenting her work at scientific conferences was Goodall's first major trial in the ravines of academic science. Leakey was getting the news around of her extraordinary discoveries and was pushing her to go public, at a time when she knew a good deal about chimps but very little of those who would be judging what she had done and how. Among her chimps she felt at home; among ethologists she was a stranger in a strange land. "My future is so ridiculous," she wrote her family shortly after the first invitations to speak. "I just squat here, chimp-like, on my rocks pulling out prickles & thorns, and laugh to think of this unknown 'Miss Goodall' who is said to be doing scientific research somewhere. Much better to be ME, I think to myself—just go out and live like the chimps—none of this scientific talk for me!!"[94] She was doing a scientist's work, yet had no sense of how to *be* one. Being dropped by Leakey's magic carpet into scientific conferences was more daunting than being dropped by boat on the edge of an unfamiliar forest. Her life had prepared her to some degree for Gombe, but not for Cambridge lecture halls and conference rooms.

Her first two invitations Goodall was able to gracefully dodge: she could not break off her work at Gombe. The next two, however, she could not escape: the first in April 1962 at the London Zoo, and the second a few weeks later in New York at the Academy of Sciences, both arranged by Leakey with all expenses paid—no excuses, no escape.[95] Most memorable at the first was the personal hostility displayed by the very eminent comparative physiologist Sir Solly Zuckerman. As chair of her session and interlocutor, he refused to take questions about her talk; so the audience, in a breach of etiquette, put questions directly to her. Then in his summary remarks he dismissed her evidence of meat eating and toolmaking as insignificant, and her work in general as a sign of the continuing dominance of "anecdote and speculation" in primate studies, intimating that it was only Goodall's personal glamour and story that got her public notice. He continued his attack afterward in interviews with reporters, even hinting that crucial observations might have been made up.[96]

Goodall could not have known what others present surely did: that Zuckerman's ostensible defense of professional standards was more a last-ditch defense of his own, outmoded, work on overcrowded captive baboons done thirty years earlier at the London Zoo. By singling her out for attack, he paid her the unintended compliment that her empirical science was the greatest threat to his science. In fact, most of those present recognized the value of Goodall's findings. As the comparative anatomist John Napier (the conference organizer) explained, "[T]his woman has redefined humanity. I asked her to the conference because Leakey said so. But it's quite extraordinary. This woman that nobody's ever heard of has appeared here on Leakey's recommendation, and look what she's been looking at." Desmond Morris likewise recalled, "She was bringing back information about chimpanzees which was, to us as students of animal behavior, tremendously exciting. We were learning things from her about chimpanzees which we couldn't have learned any other way."[97]

That was generally the pattern of Goodall's early forays into the world of professional science: ethologists accepted and valued her facts, without quite accepting her field methods or, especially, her inside stories of the science. Goodall later recalled regaling a group of ethologists with the story of how the adolescent chimp Figan, seeking to keep the senior males from hogging all the bananas, lingered after they left, only to discover that his reflexive happy food calls called the big males back—until he learned to utter the call *sub voce*, whereupon he got his bananas. Goodall expected her audience to share her delight in this revealing drama of how chimps could bend reflexive behaviors to particular situations. But instead, a chill silence filled the room, and the session chair changed the subject. Only later did she realize that the ethologists probably had been interested but dared not challenge their tribal shibboleth that anecdotes were not to be publicly countenanced as proper science.[98] (One wonders: did they murmur interest subvocally?)

It was the much the same at Cambridge, where from 1962 to 1965 Goodall worked toward her PhD degree with the ornithologist and ethologist Robert Hinde.[99] Between Goodall the forest walker and lab-raised zoologists at Cambridge there was an ambivalent relation of mutual respect and disapproval. On the one hand, Goodall's presence was a welcome opportunity for the Cambridge ethologists to move beyond captive studies of small animals to studies of free-living primates, an area that was then becoming one of the most dynamic in behavioral biology. As one of a very few active field-workers, and the most productive, Goodall was despite her unorthodoxies a valued addition to the group.[100] On the other hand, her new colleagues openly disapproved of her unorthodox methods of habituation and quasi-participation, and her stubborn attachment to a language that to regulars was unacceptably anthropomorphic. Dian Fossey, who took her degree at Cambridge a few years after Goodall, recalled that Goodall's dissertation was held up to her as "the perfect example of what not to do."[101]

However, Goodall's relation with Robert Hinde developed into one of mutual appreciation and respect for what united and divided them. (Hinde was very smart and demanding but not aggressive, and one of the few male dons who gladly took on women students.)[102] Goodall took from Hinde's instruction what was useful to her scientifically and let pass what did not. "I didn't give two hoots for what they thought," she recalled. "They were wrong, and I was right. That's why I was lucky that I never was going into any of these things for science. And as I didn't care about the Ph.D., it didn't matter. I would listen, I just wouldn't do what they said. Then I would go back to what I was doing at Gombe."[103] She did care about the PhD, of course, and about what Hinde thought. To live at Gombe she would have to learn to live in ethologists' world as well.

Hinde never tried to enforce disciplinary orthodoxy on Goodall, but devised accommodations that enabled her to do what she wanted in a way that ethologists would not dismiss as unscientific. He persuaded her to adopt methods of random sampling and numerical analysis of generic traits—not to replace but to complement her preferred mode of forest observing. And he eased her into telling tales of Chimpland in ways that did not wave the red cape of anthropomorphism before ethological bulls. Rather than saying that Fifi was "jealous," for example, she could say that "if Fifi were a human child we would say she was jealous." Loaded terms like "politician" and "social contract" were proscribed. His aim was to teach Goodall the language and demeanors of academic ethology. And despite her "somewhat truculent attitude," Goodall recalled, "I did want to learn, and I was sensible of my incredible good fortune in being admitted to Cambridge." So "[g]radually he [Hinde] was able to cloak me with at least some of the trappings of a scientist."[104]

Learning to reside in the society of ethologists was a reciprocal accommodation. Goodall persuaded Hinde that chimps were in fact individuals with personalities, feelings, and conscious agency. And Hinde persuaded her that with views and interpretations so hard to prove empirically it would pay to be circumspect. He accommodated to her way of naming and personalizing animals, and the ethological world gradually came around. When a journal editor changed her every "he" and "she" to "it" and every "who" to "which," she angrily restored her originals—and prevailed.[105] Accommodating in some matters of form, Goodall preserved the essentials of her distinctive science. So with Robert Hinde as her cicerone—her ethological David Greybeard—Goodall learned to reside and make her way in the world of high science without going restrictively native, just as she learned to live among chimps and still do science.

CONCLUDING THOUGHTS

Having crossed the species divide from human to animal science, we may pause briefly to reflect on similarities and differences. Residing was as vital to Goodall as it was to ethnographers and sociologists, but it was rather more complicated with chimps. She and her group resided in the chimps' forest home and felt at home there (but weren't really); then for a time the chimps made themselves at home with observers (but were not good neighbors); and in the end, camp and its banana "tree" became ecologically part of the chimps' forest, even as it became a permanent human enclave. It is hard to say at any point who was residing among whom. Relations between observers and observed were also less clear-cut with chimps: there could be no conversations or interviews, obviously, and no real participation beyond the simulations of the "baboon act." And though observing clues was much the same in Chimpland as in Cornerville or Kiriwina, forming conjectures from appearances about the springs and levers of social life was less secure, even with a species so like ourselves.

A distinctive and striking feature of Goodall's resident science is how much it derived from her life experience with animals and people, most notably her practice of habituation and her use of affect in divining the meaning of what she observed. In this she was by no means unique. Nels Anderson also turned his lived knowledge of hobo society, and his knack of entering into the cultures of working groups, to scientific ends. But in other cases—Bronislaw Malinowski, Robert Redfield, William Whyte—formal learning was the stronger source of resident practices. We are led to wonder what it is that causes some practitioners of resident science to draw more heavily on life experience than on academic training.

Goodall's case suggests that what accounts for the shape of resident practices is not just the subjects, or the substance of lived and learned techniques, but the order and the circumstances in which these are acquired, and how they were experienced. Which one came first, living or schooling, and in what personal and social contexts? Which one afforded the more fruitful and emotionally rewarding experience? Lacking any formal training in ethology, Goodall would naturally draw at first on practices of relating and observing in which she was already adept and which suited her new situation in Gombe. More important, however, is how productive as empirical science her life practices proved to be, almost from the start; and how emotionally rewarding were the situations in which life practices became science as well. She loved her new forest home and was happiest there, and made significant discoveries within months of her arrival, doing what to her came naturally. Such experiences could only confirm her in unorthodox practices of habituation and strengthen her skepticism of the formal training that she did not have but knew she would one day have to endure. From the circumstances and timing of her life history it is quite understandable that she would resist orthodox ethological practices that seemed to her less revealing of chimp society and were far less satisfying personally. She persisted in her distinctive ways of resident observing for a lifetime, even as she learned to reside and make her way in the culture of professional science.

That Goodall would experience residing in Cambridge as a greater challenge than residing in the Gombe forest was an inversion of the usual pattern of scientific lives. Learning the ways of academic society more typically comes first, and learning to reside in nature is the challenge of the new and untried—as it was for the hypereducated Malinowski. For those whose formal training and socialization follow a worldly apprenticeship—especially a quickly successful and emotionally rewarding apprenticeship like Goodall's—the schooling part is the challenge of the new. It is common for individuals' life experiences to have some effects on their subsequent science—it was true more or less for all the main figures in this book. It is not so common for a person to be a productive investigator before acquiring formal training and credentials, as Goodall did. Hers is a case of what we might label an "inverted" life history or career.

The causes of such inversions in life stories are likely always to be singular and idiosyncratic, as they were in Goodall's case. But her particular case alerts us to the possibility that learning to reside in an academic or professional milieu is potentially significant in any case and should not be taken for granted as simply "how it's always been done." Resident science is a two-faced occupation: practitioners reside both in some local subject group and in a cosmopolitan scientific community of peers and judges. So we might expect to find cases of inverted careers more commonly in sciences done intensively in situ, because for some it may be the subject group that is the more easily engaged. It was in fact Goodall's case that made me take a second look at Malinowski's academic training and opened my eyes to his laboratory dead end and its possible effects on his subsequent ethnography. Variants on this pattern of career inversion will figure in the life histories of wildlife ecologists Paul Errington and Herbert Stoddard, coming up next.

CHAPTER 6

WILDLIFE ECOLOGY

THREE LIFE HISTORIES

The subject of this concluding case is not another dramatic celebrity species but a low-profile homey species of game bird, the bobwhite quail. And it poses the question: If resident science is defined by coresidence and an interactive relation of some sort between observer and observed, can the concept apply to a creature whose relation to humankind is mainly to be sport and food, and whose response to human presence is to bolt or take cover? There's no habituating or coresiding here. The situation is further complicated by bobwhites' preference for living in the human environments of farms and woodlots, making them participants in a human as well as a natural ecology. The case thus calls for a more expansive and capacious concept of residence, and, as we will see, for an enlarged view of inverted life histories.

The case comprises three closely connected projects from the late 1920s and 1930s.[1] Herbert Stoddard investigated bobwhites in private quail preserves in southern Georgia and Florida, the core of the species' range, while Paul Errington worked at their northern limit in rural south Wisconsin. Aldo Leopold, who was close to both Stoddard and Errington, drew upon their empirical work for his celebrated environmental ethic of land health. The three principals shared an early and abiding love of hunting and outdoor life and were dedicated environmentalists; they nonetheless took somewhat different approaches to their subject. Stoddard pursued a comprehensive life history of bobwhites to the practical end of ensuring abundant game for fall hunts, whereas Errington focused exclusively on the ecology of "natural" (that is, not human) predation. Leopold was instrumental in organizing the two projects, but did not take part in the actual fieldwork and was interested above all in the larger implications of its empirical results. His philosophy of land health was in effect a theory of human and wildlife coresidence. These differences, I will suggest, derived from the life histories of the three, especially from differences in the ordering and intensity of their life activities and formal training.

This conjuncture of three life paths in the bobwhite projects was no chance event but a product of a common context of public concern about game species. Sport hunters and fish and game officials were at the time (1910–20s) worried by the steady decline in the abundance of game species, especially waterfowl and other game birds, and deer. Time-honored remedies were ongoing: large government programs of annual game stocking, tighter regulation of hunting seasons and bag limits, and organized campaigns of exterminating competing predators—coyotes and wolves, cougars, winged raptors (hawks, eagles, owls)—led by no less a power than the US Bureau of Biological Survey. Yet game continued to decline, and by the mid-1920s there was a growing sense, especially among academic zoologists and conservationists, that the usual remedies were misguided. Annual stocking was no long-term cure, and eliminating predators resulted in population explosions of rodent and herbivore species that destroyed the environments on which all resident species, including humans, depended. As a countercampaign against wholesale slaughter of predators gained momentum, wildlife and land-use politics became intense and partisan. It was in that situation that the felt need for a basic science of game and wildlife became urgent and took concrete shape in undertakings like the bobwhite projects.[2]

The thinking in reform circles was situating: healthy populations of game animals were best assured not by artificial short-term interventions, but by maintaining intact, healthy environments in which resident wildlife could take care of themselves and thrive. That was the idea behind the bobwhite projects, as well as Leopold's 1933 book on game management—the constitution or bible of the nascent profession—and his environmental ethic.[3] Of course, long-term environmental solutions were not easily sold to ranchers who lost stock to predators and sport hunters who wanted only more game right now. And environmentalists were hampered by the lack of credible empirical evidence. Beyond the abstract knowledge of zoology and the practical experience of hunters, farmers, game wardens, and naturalists, not a lot was known for certain about the hidden lives of wildlife species. Empirical investigations in situ were thus crucial to creating a science of wildlife. That was the context in which the bobwhite projects were hatched.

It mattered that wildlife studies took shape within an economy and culture of subsistence and sport hunting.[4] Although the term "wildlife" in time came to mean any free-living species, in the formative years of wildlife studies it was more likely to mean game, especially ungulates and game birds. Game animals contingently stood in for all free-living species. Another consequence of that context was that the science of wildlife was from the start conjoined with the professional and managerial side of the subject. It was what Lynn Nyhart has termed, following the German usage, a *Kunde*.[5] (The English word "studies" does the same work, as in "science and technology studies.") The distinction between knowledge and know-how was thus less marked in wildlife studies than it was in other sciences, and that left an open field for individuals with various backgrounds and experience to practice wildlife science and management without, so to speak, a license.

The connection with a deep culture of hunting and outdoor recreation meant that in wildlife studies formal education was not (yet) the obligatory path to professional careers that it was in established sciences. (Leopold's students would be the first credentialed game managers.) Personal experience and knowledge of outdoor life was an acceptable path to professional careers, and love of nature an acceptable attitude for science. This was a window of opportunity for people like Stoddard, who had only a grade-school education, and Errington, who came late and somewhat reluctantly to higher education. And their worldly perceptions and practices became foundational to the new science. Errington turned the practices of hunters and trappers—close observing, tracking, reading sign—to scientific ends. Stoddard's practices of land management were refined forms of time-honored practices of local farmers and foragers. And Leopold's conviction that small farmers were potentially the best environmentalists reflected his lifelong engagement with his region's mosaic landscape and residents.

STODDARD: RED HILLS

Colinus virginianus, the bobwhite quail, is a sedentary ground-dwelling bird that seldom strays too far from home.[6] Strong in flight only in sprints when flushed, bobwhites do not migrate and generally live out their lives within a few miles of where they were hatched. In spring they disperse in nesting pairs, then congregate for the winter in coveys of six to twenty birds. An early-succession species, bobwhites evolved to live in patches opened in woodlands by fire or storms, or in open woodlands maintained by fire in an early-succession state—like the longleaf or yellow pine savannahs of the southeastern United States. Bobwhites eat a varied diet of seeds, berries, and insects, but favor early-succession grasses and legumes. They avoid dense brush and understory, in which ground-dwelling predators lurk unseen, as well as open pasture and croplands, which afford no cover from winged raptors. They are regular in their routines, moving predictably among feeding, roosting, and dusting sites.

Despite their (understandable) wariness of humankind, bobwhites thrive in areas of small-scale family farms, where a mosaic of small cornfields, pasture, and woodlots with weedy fencerows and brushy wastes affords abundant food and handy cover, and where seasonal plowing and mowing maintain the early-succession ecology. Their populations boomed when old-growth forests were cleared for farming. Small farms afford other amenities of bobwhite life: cornfields and corncribs for winter food; farm outbuildings, machinery, and abandoned car hulks for emergency cover and nesting spots; and hens' nests to occupy (an endearing though unhygienic habit). Rural folk regarded bobwhites with affection, not just as good to eat but good to think about. They were exemplars of family values—fiercely protective nesters—and of perseverance in disruptive situations, a resonant association in the post–Civil War South.[7] To human residents, bobwhites were steady and welcome cohabitants: "an ally of the farmer" with "a strong hold on our affections for aesthetic reasons," as Herbert Stoddard put it.[8] Like their human cohabitants, bobwhites were settled residents of native ground, and thus good subjects for resident study.

The site of Stoddard's investigation (1924–29), in the Red Hills of southern Georgia and northwest Florida, was in the core of bobwhites' range, with a rich human and natural ecology. Once the dominant forest in some sixty-five million acres of the coastal plain and piedmont from Virginia to east Texas, longleaf pine depended on regular fires that left young longleaf unharmed and killed off the faster-growing loblolly and slash pine and deciduous species that would otherwise have become the climax forest. The longleaf forest had been largely destroyed by clear-cut logging in the notorious "industrial cut" of the early twentieth century. However, owing to its unusual topography and history, the Red Hills remained an island refuge of some 200,000 acres of this distinctive natural and human ecology.[9]

Although cotton had been grown here and there in the Red Hills, the rolling terrain was unsuited for plantation agriculture, and farm clearings remained small and dispersed in a savannah woodland kept open by a custom of annual burning of undergrowth begun by Indian hunters to foster game species and ease travel. This ecological regimen changed little in the post–Civil War devolution of plantations to small-scale tenantry, and in the subsequent rise of a resort economy as the Red Hills became a winter destination for well-to-do northerners seeking a healthy rural landscape that was tame but with a touch of the unkempt and culturally exotic. As quail hunting became an important tourist draw, local landowners took the opportunity to lease, and then sell, land to wealthy northern sportsmen; and small parcels gradually coalesced into huge preserves of up to ten thousand acres or more. Tenant farmers were not evicted: they were essential to the natural and human ecology of sport hunting, keeping the woodlands open and serving as estate stewards and guides. The old practice of annual burning was continued, to ensure a well-fed population of quail for the fall hunts. And the local custom of hunting small mammals (raccoons, foxes, possums) had the useful side effect of keeping these egg and chick eaters in check. The Red Hills was thus a resident, working landscape: an "inner frontier" accessible to humankind but not overrun and overwhelmed.[10] In this distinctive natural and human ecology, *C. virginianus* was the keystone species. By the early 1920s, however, it was also a declining species.

The Red Hills Cooperative Quail Investigation was initiated by a group of local landowners who were alarmed by the decline of quail and sought the aid of the US Bureau of Biological Survey in diagnosing causes and devising remedies. Survey scientists brought expertise and oversight, while landowners paid salaries and expenses and granted unfettered access to their properties. The Quail Investigation was an unusual project, both in the sheer size and ecological diversity of available research sites (200,000 acres in all) and in the landowners' unqualified support for a scientific approach rather than the usual game stocking and predator roundups. They may have been primed by S. Prentiss Baldwin, a Cleveland businessman and accomplished amateur ornithologist, who in the 1910s had developed the modern practice of banding adult birds netted en masse, in Ohio and at Inwood Plantation, his Red Hills estate.[11] As a member of the sportsmen's own social set, Baldwin could speak with social as well as scientific authority for a scientific approach. It was likely through banding that the Biological Survey found Herbert Stoddard, an early and active convert to Baldwin's methods.[12]

Stoddard was not an obvious choice to head a scientific investigation. Born in Rockford, Illinois, in 1889, he was an assistant taxidermist at the Milwaukee Public Museum in 1923 when the invitation arrived out of the blue to head the Red Hills investigation. Though poor in formal schooling, he was rich in experience of life in the woods: first as a boy in the outback of east-central Florida, where his family had moved in 1893, drawn by booster dreams of orange groves; and then, when those fantasies collapsed, back in Rockford and the wildish country around Prairie du Sac, Wisconsin, where the Stoddard family had been pioneer farmers. There Herb pursued a mixed life of farming and trapping and a growing business in taxidermy, which led to positions as assistant taxidermist in the Milwaukee Public Museum (1910–13, 1920–23) and the Field Museum of Natural History in Chicago (1913–20).[13]

Although his museum experience had made Stoddard, in his words, "an ornithologist of sorts," it was not his rather spotty science that attracted the organizers of the Quail Investigation, but rather his expert woodsmanship and firsthand knowledge of animals in their home environments.[14] Stoddard's taxidermy did give pause. Though taxidermy had once been an accepted path to museum careers, by the 1920s credentialed scientist-curators had gained control of building collections and dioramas, demoting taxidermy from protoscience to technical occupation.[15] The Biological Survey's Waldo McAtee worried that Stoddard's occupation and lack of education would compromise his credibility with the Red Hills landowners. In the event, Stoddard was embraced by local sportsmen, estate managers, and tenant farmers as someone whose life experience and outlook were very like their own.

For Stoddard himself his lack of formal education was "a continual source of embarrassment." As he became an authority in wildlife studies, colleagues would assume that he had credentials to match and would ask awkward questions. These he learned to parry, by drawing a distinction between *schooling* and *education*: of schooling he had little, but was richly educated by the varied experiences of his life. That worldly education, in the words of historian Albert Way, "most closely resembled a series of apprenticeships."[16] And those practical apprenticeships were good preparation for the kind of wildlife science he wished to do. So Stoddard took up residence at Sherwood Plantation, the thousand-acre retreat of the principal landowner and patron, "Colonel" Lewis Thompson.[17]

Stoddard called his bobwhite study a "life history," a capacious term for a descriptive account of an animal species that included some mix of morphology, taxonomy, physiology, and behavior, as well as ecology, relations with humankind, and folklore. Stoddard's version of life history emphasized topics essential to bobwhites' ability to survive and multiply. The plan he presented to Waldo McAtee after six months of exploratory fieldwork focused on the life-and-death factors of provisioning and mortality.[18] What were bobwhites' staple foods, and what determined their abundance? What were the chief causes of bobwhite deaths, and how did the birds evade predators and use their intimate knowledge of their home environments to live and raise young?

Stoddard's book, *The Bobwhite Quail*, gives a synoptic view of his life-history approach.[19] Reproduction is prominently first: courtship and mating, nests and nesting, egg laying and incubation, hatching, rearing and brooding, learning adult habits, forming fall coveys, with a bit on the ornithological topics of morphological variation and sex ratio. Next come vital life functions: diet and foraging behaviors, diseases and pathology, and predation, especially of eggs and chicks in nests. The book concludes with chapters on techniques of improving quail environment by planting, plowing, controlled burning, and modest predator control. The idea was to maintain large and healthy populations of birds not by annual stocking and predator roundups, but by maintaining healthy environments, and by the time-honored practices of keeping nature in an early-succession stage, in which bobwhites thrived. Stoddard's book became a model for a generation of game and wildlife scientists and managers.[20]

Stoddard's field practice likewise combined science and use. He relied on scientific methods—close observing of animals in situ, plus experimental work with captive animals in controlled settings—but also made free use of residents' vernacular customs and intimate knowledge of bobwhite haunts. Estate managers and tenant farmers found and helped monitor nest sites, and served as guides in unfamiliar terrain. Most important, Stoddard kept always in mind the total environment in which bobwhites lived and died. It was not just food and predators that mattered, as many believed, but the birds' total situation. In Albert Way's words, "[H]e would never approach the study of ecology in abstract theoretical terms. Instead, his explanatory device would become the land itself."[21] It was an ecology of coresidence. Although much of the fieldwork was done at Baldwin's large Inwood Plantation or nearby, Stoddard also sampled the entire region's varied topography and ecology, to get a sense of the range of variation in quail habitat and behavior. "The complete life history of the bobwhite can never be written," he cautioned readers, because "exceptional behavior is common in exceptional conditions."[22] One is reminded of Jane Goodall's combined use of wide-ranging forest walks and camp provisioning. Stoddard's life-history science was a situated and a situating science: a science of context, observing, and reasoning in situ.

In devising techniques of land management, Stoddard again drew equally on science and on residents' practices of farming and foraging. Practical interventions did not spoil the science but were themselves "natural" experiments, in which the effects of changes in the environment on bobwhite behavior and survival could be observed. The results of his investigations confirmed the wisdom of some everyday beliefs and practices, while debunking others. The time-honored custom of controlled burning was confirmed. Stoddard was initially cautious about this practice, thinking that bobwhites could not possibly live in bare, blackened burns. Yet the birds not only stayed but favored burns for foraging—where, he then noticed, their preferred food plants grew most abundantly. Experiments with captive birds confirmed their preference for the weeds and grasses that follow fire.[23] This time-honored practice, if done at the right season and in the right way, was an easier and more effective way of maintaining quail habitat than sowing selected food plants. Likewise with periodic plowing of fallow fields to destroy brush and release forbs—another common activity of tenant farmers. Experiments with this practice showed exactly when and how plowing should be done to encourage quail food plants without damaging the total environment.

It was a different story with the equally time-honored custom of killing all hawks and owls on sight, in the belief that they were responsible for declines in bobwhite populations. Empirical study proved that these indiscriminate interventions were in fact not just ineffective but counterproductive. Among raptors only Cooper's and sharp-shinned hawks were serious threats to quail, owing to their exceptional stealth and speed of attack (hence their common name of blue darters). Other hawks and all species of owls were shown to be modest eaters of quail, but prodigious consumers of the rodents and other "varmints" that fancied quail eggs and chicks. Killing them was killing hunters' friends. In contrast, the local custom of hunting raccoons, possums, and skunks—all quail eaters—for meat and pelts demonstrably boosted quail populations.[24]

The "natural experiments" of residents' everyday activities, tested and refined by experiments of his own, became central elements of Stoddard's game science and management. To once more quote Albert Way, "Rather than forcing a cultural landscape to fit an abstract set of scientific principles, he set out to mold a system of management from the region's cultural and environmental past. In addition to being a land of both respite and work, the circumscribed environment of the quail preserves was now an open-ended scientific laboratory."[25] Stoddard had only to observe customary beliefs and practices and then select or fine-tune them for best effect. Promoting controlled burning and opposing raptor roundups became major preoccupations of his career in game management.[26]

The practices of Stoddard's science were likewise mixed. Observing free-living birds in nature was his preferred mode, but he also observed captive animals where that seemed the best or only way to get the facts. Just so much could be done in situ. Bobwhites were adept at keeping a screen of vegetation between themselves and would-be watchers—their skill in evasion was in part what made them such a valued game bird, testing hunters' own skills. But observers' patience was also tested. So they welcomed the rare occasions when quail were feeding near a house and could be watched through a window. Stoddard also discovered that coveys could be followed and observed from a car with curtains drawn. But courtship and mating displays were very private and could be observed only with captive birds in enclosures stocked with native vegetation to make courting couples feel at home. Field data were "more or less fragmentary," Stoddard confessed, and served mainly to check if captive birds behaved the same as birds in situ (for the most part they did).[27]

Nesting behaviors, in contrast, were best observed in situ, because quail on nests tended to stay put and defend them. (Bobwhite parenting was, in Stoddard's admiring words, "beyond reproach.") Nests were always well camouflaged but could be found by those who knew the ground, and observed from blinds. All told some 2,160 visits were made to 602 nests and daily logs made of environmental conditions, techniques of camouflage, numbers of eggs laid, numbers of nest failures and second tries, and details of incubation and care of hatchlings: a full and intimate picture of that crucial stage of bobwhite life history. Actual laying of eggs was rarely witnessed (some dozen times in five years); and though hatching could be observed in nests, the better way was to take eggs in hand and watch the chicks break out. Observing nesting pairs in captivity was useless, since enclosure disrupted normal parenting.[28]

Nests were also the chief sites of quail mortality. Of the 602 nests with eggs that were observed, about 64 percent failed, for reasons that included weather extremes and disturbance by humans or farmyard animals (cattle, poultry). Natural predators (mammals, snakes, ants) accounted for some 37 percent of failures. In most cases cause of death could be inferred either from circumstantial evidence or physical remains. Weather could be inferred as the likely cause of failure of nests in low wet spots in rainy spells, or in dry hot spots in periods of heat and drought. Farm activities could safely be blamed in the case of nests at the ends of corn or cotton rows. Egg and chick eaters were rarely caught red-handed, but abandoned nests (119 of the 602) were rich in clues to those who knew to see them.[29]

The condition of broken eggs was especially revealing of the identity of egg eaters. Skunks bit off the tops of eggs in the nest, then licked out the contents. Cotton rats left shells with clean-cut edges along their runways, while weasels removed eggs one at a time to a nearby tangle and bit off the tops. Possums also removed eggs singly but swallowed them whole. Dogs with their messy open-mouth eating left a scatter of shell bits, while turkeys tore off tops and tossed them to one side. Snakes were implicated by long scat with rows of egg membranes. Eggs with tiny skeletons inside implicated ants. It was harder to ascertain what killed hatchlings outside of nests; the tiny corpses disappeared quickly (as experiments confirmed). But quail bones in scat or regurgitated owl pellets, or in the stomach contents of raptors, were telling clues of whodunit for those who were literate in the material language of bits and bones.[30]

Much of the forensics of "reading sign" would be known from experience hunting and trapping—following trails, analyzing kill sites—and diagnostics not used in hunting were discovered by experiment. The signature styles of egg eaters, for example, were revealed when captive predators were given eggs and watched. Diet preferences were likewise learned by presenting captive animals with a choice of prey species.[31] Wildlife science as it developed in the Quail Investigation was thus a mixed practice of intensive in situ observing of nesting and foraging in the quail preserve, backyard experiments at Stoddard's home base of Sherwood Plantation, and wider observing of variations in quail behavior in the greater backyard of the Red Hills.

STODDARD: LIFE AND SCIENCE

The distinctive features of Stoddard's science—close observing in context, a preference for life history over biological ornithology, active relations with local residents—were deeply rooted in Stoddard's previous life activities, as he himself was well aware. Looking back, he saw that "almost everything I had done so far in my life, from my boyhood activities in the Florida pinelands, had been good training for life-history investigations of birds."[32] Albert Way noted that every stage of Stoddard's life contributed something to his distinctive science: "his background as a curious kid, his informal training as an apprentice taxidermist, and his coming of age as a professional ornithologist and fieldworker in natural history museums."[33] His wildlife science was a more or less direct continuation of his everyday activities as outdoorsman, trapper, and taxidermist.

In all his early experience with animals in nature Stoddard learned most from people who knew the ways of wildlife from their practical occupations. His experience of animals was thus always of their relations with humankind: he drew no sharp line between animal and human ecology. When the killing freeze of the Stoddards' orange grove made schooling an unaffordable luxury, young Herb was left largely on the loose in a backwoods that for a nature-loving boy was a paradise. However, the curious kid did not just play around in his subtropical paradise. He wanted to learn about it and its plants and animals, and realized that the best way to learn was to attach himself to longtime residents who knew the place and its animal residents intimately. From a Mr. Barber, a retired surveyor and keen hunter and amateur naturalist, Herb learned the art of trapping and making animal skins, which taught him much about animals' habits and habitats and earned welcome additions to his mother's meager income.[34]

However, his most valuable education in the Florida outback came from the cattlemen who ran their stock through the grazing commons of savannah woodland, prairie openings, and swamp hammocks. The area's leading cattleman, Gaston Jacobs, was a neighbor, and Herb began hanging around his cattle pens doing odd jobs for free, then as a cattle boy assisting with the herding and pasturing. His constant companion was Gaston's son, George, who was a few years older and an experienced backwoodsman and willing teacher. In the mosaic world of island prairies and patch-burned woodland, cattle were grazed wherever there was a patch of forage; and the improvisations of commons herding gave George and Herb ample opportunities to look around. Having settled the herd to graze, they were at leisure to shoot, trap, fish, forage for berries and wild cabbage, discover birds' nests and bee trees, and just explore and satisfy their curiosity about animals they saw going about their everyday lives. When they ran out of provisions, they, too, foraged. And when the cattle wandered off, as they often did, the boys had to imagine where they might have gone, thereby acquiring an intimate knowledge—an animal's knowledge, one might say—of the environment.

Herb learned how not to get lost in a bewildering terrain—a skill vital to a cattleman's identity and self-respect—and how to live off the land and avoid its very real dangers. With George he gained a living knowledge of how animals used their environment to survive, and he had his first lessons in ecology, as when George stopped him from killing a species of snake that ate varmints. "Never have I relished life more," he recalled, "than I did during the years I spent herding cattle from 1896 to 1900." He also had his first experience with woodland burning when, having watched cattlemen burn patches of overgrown savannah to improve the forage, he thought he would do the same with the matted grass around his family's home. Alert neighbors saved the house (just barely), but the fire burned for three days over many miles. (Old-time residents were unconcerned, and no real damage was done.)[35]

Young Herb managed largely to evade schooling, but as an apprentice cattleman he acquired an education in animal life that was incomparably rich. "[N]o schooling or advantages," he recalled, "could have been more valuable to me. I firmly believe that all experiences become a part of a man. Certainly my years in the southern pinelands . . . were invaluable to me in my later years as ornithologist, ecologist, and wildlife researcher and manager."[36] We can see in that early education Stoddard's future preference for a lived science, and his conviction that woodsmanship was vital to wildlife science and management and should be part of professional training. Albert Way writes that "these 'cattle hunters' were Stoddard's first model for the informed land manager he would become."[37] Yet occupational prospects in Florida were limited. Though rich in nature, its human culture afforded no points of entry into the wider world of natural history. Herb knew he wanted to be a taxidermist; but there was no one there to teach him (Mr. Barber's knowledge of the art being limited). Though leaving Florida when his family returned to Wisconsin, in 1900, was for Herb an exile, the upper Midwest did afford a wider scope for his expanding ambitions.

The landscape of the Stoddard family's home region around Prairie du Sac was, like central Florida, an inner frontier, with a distinctive residential history. Part of the Driftless Area, which had escaped the scraping and burying of glaciation, it was a dissected landscape of flat-topped hills (remnants of an ancient peneplain) and steep-cut stream valleys. Unsuited to mechanized farming, the place survived as an island of woods and small farms in a treeless Corn Belt sea. A poor place for farmers, it was a fine one for woodsmen and hunters. It also had a vigorous regional culture of woodsman-naturalists that was tied into the region's established system of high schools, small colleges, and amateur societies, in which nature-going and natural history were popular pursuits.[38] It was a world both rich in nature and close to towns and cities—Madison, Milwaukee, Chicago—with excellent museums and universities and a cosmopolitan culture of civic science that was open and inviting to an aspiring young trapper-taxidermist.

The farm where Stoddard began work in 1905 belonged to an immigrant family named Wagner and afforded ample opportunities to be out in the woods with the Wagners' son Arthur quarrying, timbering, trapping and hunting, and observing wildlife. In 1906 Stoddard persuaded a local naturalist and taxidermist, Ed Ochsner, to teach him the art; and for the next four years he commuted between the Wagner farm and Ochsner's shop in Prairie du Sac, where he served as assistant and apprentice, talked endlessly about birds and animals, and listened enthralled to stories of the celebrated taxidermists of the Milwaukee and Chicago museums, George Shrosbree and Carl Akeley. He was introduced to other local hunter-naturalists, most notably Albert Gastrow, a farmer and skilled woodsman, tracker, and hunter with a deep knowledge of local animals. Stoddard became the youngest member of this circle of woodsmen-naturalists—"Ochsner, Gastrow, and Stoddard. Birds of a feather were we." And through Ochsner he gained entry to the world of museum taxidermy and ornithology, in the Milwaukee Public Museum.[39]

Stoddard's main occupation as a museum taxidermist was building small mounts of common birds that circulated to great acclaim in the state's public schools.[40] It was a less prestigious taxidermic role than building scientific study collections or dioramas, but it suited Stoddard. The exhibits illustrated birds' life histories, which he knew well from his own experience in the woods. And because the birds were local, not exotic, species, he could do his own collecting, in nearby urban hinterlands (then still rich in wildlife) and especially in the area that he knew and loved around Prairie du Sac and the Baraboo Hills. Using his local knowledge and contacts, he gathered hard-to-get material for a series of life-history groups of nesting raptors with their young. At the Wagner farm he climbed high trees to collect eggs and hatchlings, and cut out sections of trees with nests. For this arduous work Stoddard enlisted the aid of another local woodsman, Bert Laws, who with his wife, Annie, worked a farm in a remote site on the Wisconsin River. "It was then a wild and glorious country," Stoddard recalled, "accessible only by a tortuous sandy road from Sauk City or Mazomanie, or by boat along the river." Unlettered and uninterested in reading and writing, Laws much preferred naturalizing to farming and had acquired an incomparable knowledge of local wildlife. He would drop everything to help Stoddard, who returned the favor by helping with chores and milking.[41]

It was Laws's example that taught Stoddard the difference between schooling and education and convinced him of the need for practical woodsmen in a science that depended on their know-how. "I have long been convinced," he later wrote, "that outstanding woodsmen-naturalists, exemplified in such men as Gastrow, Ed Ochsner, and Bert Laws, are indispensable to the success of investigations of wary birds and mammals. All too often scientists are conspicuously lacking in woodsmanship."[42] Stoddard thus entered the world of museum science without leaving everyday naturalizing behind.

But as Stoddard grew as a museum ornithologist, that happy accommodation of life and science became harder to maintain. The life of a professional zoologist, he gradually realized, was not one he wanted to live. A decisive experience was the Milwaukee museum's 1922 expedition to the world-famous seabird nesting site of Bonaventure island in the Gulf of St. Lawrence, to gather taxonomic specimens and material for an elaborate "Bird Rock" diorama. It was Stoddard's first taste of an organized museum expedition: collecting specimens en masse for study collections together with elaborate field notes, plus accessory materials for the diorama's foreground; and making photographs and films of the site for the illusionistic backdrop. Rappelling down cliffs to collect and observe was exciting. But back home with thousands of specimens to work up and classify and a complicated diorama to construct, Stoddard realized that the higher he rose in a museum career, the more his life would be urban and indoors.[43] Moving up meant staying in. What he wanted was an occupation that combined learning and outdoor life, like his apprenticeships in Florida and the Baraboo Hills; and the Red Hills quail project seemed just that. Though it was a temporary assignment, he sensed that he would have a future in game science and management. "A new profession, wildlife management, was being born," he recalled, "and further studies of game species would almost inevitably follow this one."[44]

Serendipitously, the Red Hills eventually became the Stoddards' home. After the Quail Investigation was finished, Stoddard stayed on as a wildlife expert with the Biological Survey, finishing his bobwhite book and doing survey business—the path of least resistance. His boss, Waldo

McAtee, hoped the arrangement would be permanent; but Stoddard's year in Washington (1929–30) only confirmed his dislike of cities and of the life of a government expert working on projects in which he had no expertise or interest. He wanted to follow up his study of quail with life histories of other game birds—western quail, turkeys, and prairie chickens were favorites—in a rural environment as agreeable as the Red Hills. Getting wind of his desire to leave the survey, Colonel Thompson offered to give the Stoddards Sherwood Plantation, if Herbert would set up practice as a private game consultant and devote himself to helping landowners make their properties hospitable to game birds. Sherwood, Stoddard thought, "could provide enough ornithological projects to keep me fully occupied for the rest of my life."[45] He thus became once more an observing resident and resident observer—a role not unlike those of his early apprenticeships.

Meanwhile, scientific field studies of other game-bird species were getting under way in other parts of the country.

ERRINGTON: A SITUATING SCIENCE

Paul Errington's bobwhite investigation, which began in 1929 as the Red Hills project was winding down, was part of a larger project inaugurated in 1928 by the Sporting Arms and Ammunition Manufacturers' Institute (SAAMI), a trade association created in 1926 to set standards of gun performance and safety and to attend to the industry's long-term future. The steady decline in numbers of game birds was, obviously, a threat to the future of sport hunting and thus to makers of sporting arms. Especially worrisome to SAAMI was the lack of objective measures of the decline and of scientific understanding of its cause. Nor was there a body of wildlife professionals who could be called upon to gather data and devise remedies. So in 1928 SAAMI hired Aldo Leopold, who had just resigned from the US Forest Service to become an independent game consultant, to carry out a quantitative game survey—which confirmed the general decline and gave it the numerical authority of science—and to set up a program of graduate fellowships in game management.[46]

The plan was to produce empirical studies of four game species as models for further research, and at the same time to train a founding cohort of wildlife professionals. Stoddard was borrowed from the Biological Survey to help select the first four fellows, and in October 1928 he and Leopold made a three-week tour of seven state universities, from Ohio to Arizona. While Leopold assessed campus personnel and facilities, Stoddard scouted nearby sites for field research. The two men quickly became close friends and coconspirators in the cause of wildlife study.[47]

There were several reasons for choosing bobwhite quail and the University of Wisconsin for one of the SAAMI fellowships. Leopold was angling to become the first professor of game management, at the University of Wisconsin. (He was appointed in 1933.) And southern Wisconsin was ecologically an ideal place for a second bobwhite study that would complement Stoddard's in the species' core range. At the northern limit of their range, quail populations grew and expanded in periods of mild weather, then declined and retreated when cold and snow returned. The edge of habitability thus seemed an ideal situation for observing cause and effect between environmental conditions and abundance—natural experiments. And with small numbers of birds a single worker could count and monitor the daily lives and fates of every one. Another plus was Paul Errington, then a beginning graduate student in zoology, who as a subsistence hunter and trapper had the practical experience in woodsmanship that Leopold and Stoddard valued—plus independence and initiative, and a very definite idea of what to do.[48]

Whereas Stoddard sought a complete life history, Errington was interested solely in the ecology of predation on adult birds: the life-and-death encounters of predators and prey, their strategies of attack and defense, and the environmental circumstances that made attacks succeed or fail. He studied only winter mortality, as the measure of predation, and did nothing with nest and brood mortality. The fall density of birds was of no interest to him, as it was to Stoddard in the southern quail-hunting preserves. For the same reasons he refused steadfastly to intervene in the lives of quail, whether by feeding in weather emergencies of deep cold or snow, or by planting food and cover plants. Although he was an avid hunter, as an ecologist it was not his business to improve the hunt. In a study of the ecology of predation in natural conditions, human interventions of any kind could only spoil the science. "When [quail] were dying," he wrote, "I usually let them die, just as quail always have died, without either relieving or aggravating their troubles."[49] Disinterestedness was a rooted scientific norm, though I will argue that it derived less from Errington's scientific training than from his earlier experience as a subsistence trapper.

Errington did all the fieldwork himself from 1929 to 1932, when he received his PhD. He then left Madison for a post as research assistant professor in wildlife ecology at the Iowa State University School of Agriculture in Ames. There, again with Leopold's help, he undertook a second study of quail predation, plus a large parallel field study of muskrats and minks, their chief predators. The Wisconsin study continued under Leopold's supervision, with the fieldwork delegated to Stoddard's old friend Albert Gastrow, who was retired and living in town but knew the study area intimately as a lifelong resident. Errington was for most of the project a resident only virtually. He made occasional visits but mostly analyzed Gastrow's field notes and wrote up results for publication. This arrangement continued until the early 1940s, when Leopold began to use the project for graduate students' dissertation projects. The project closed in the late 1940s, when Leopold died, and funding, always precarious, finally ran out. (Errington and Leopold had often paid Gastrow out of their own pockets.)[50]

Errington worked several study sites around Madison, but mainly an area of 4,500 acres of small family farms near Prairie du Sac.[51] This site had the advantage of being enclosed on two sides by the Wisconsin River and on the other two by open cropland—a death trap for lost or wandering birds—so that resident bobwhites tended to stay put. The site was in effect a natural laboratory for quantitative study of a stable population. The area's human ecology was also favorable. Northern farms did not require controlled burning or long fallowing, as farms did in the Red Hills. Except for occasional fits of debrushing, farmers were content to leave field edges and woodlots unmanaged and untidy—good quail habitat. And agricultural practices changed little over time, which reduced the contingent noise in natural ups and downs of population. In addition, low population densities rarely warranted an open hunting season on bobwhites (only two were declared in the twenty years of study), leaving populations largely undisturbed for scientific study of "natural" predation. As Errington noted, marginal habitat provided "by far the best outdoor laboratory I know of for getting closer to what actually 'makes the wheels go around.' "[52]

Human residents of the study area were nonetheless essential to Errington's project. Since the land was privately owned, researchers were legally trespassers and depended for access on residents' interest and goodwill. The first thing Stoddard did for Errington was to take him around and introduce him to his old friends and neighbors, most notably Albert Gastrow.[53] As Errington later advised a beginning ecologist, study sites should be chosen not just for their natural advantages but also for the prospects "for sustained interest and cooperation on the part of human residents."[54] His working maps of the study area chart property boundaries as well as bobwhite covey territories.[55] Although he never lived on site, he had a resident's knowledge of its human and animal ecology. He was in effect a commuting winter resident, going about his daily occupation of science much as resident farmers went about theirs.

To measure predation Errington needed to know two numbers: total winter losses, and losses specifically to predators. That required precise counts of population in November and April, and an accounting of the exact causes of individual deaths—ideally of *every* death. (In practice he got about half.) Both census and mortality forensics were exacting and arduous practices. The most accurate method of census was the tracking count: observing the tracks made by coveys as they moved about, and reckoning from these the number of birds present. This technique depended on the bobwhites' predictable domestic routines. Errington could quite easily discover coveys' habitual roosts and their favored paths between these and feeding spots—quail "highways" he called them. But good track counts depended on good conditions: a light snow was best, but dust or mud would do. Though simple in principle, in practice tracking counts required seasoned judgment. Covey tracks were often hard to read: birds walked single file, crisscrossed, backtracked, or walked and flew, thus vanishing and reappearing from the trails in the snow. And coveys kept reshuffling, so had to be continuously monitored and updated. It was very much a resident practice. Typically Errington made counts every week or so, and every day in periods of reshuffling or weather emergency.[56]

Discovering the causes of mortality was even more exacting. Since observers were seldom lucky enough to witness actual deaths, causes were typically inferred from circumstantial evidence or from physical remains at kill sites. Corpses of emaciated or frozen birds were unambiguous clues to the cause of death, and marks of fluttery flight in the snow were signs of feeble takeoff and incipient starvation. Fit and well-fed quail took off explosively. The signs of deaths by predation were richer, but reading them required knowledge of the distinctive killing and eating habits of predator species. Feathers arranged compactly ruled against a hawk or owl, which pluck and toss. Fragments of crushed bone testified to small mammals, since hawks can't break larger bones, and owls bolt chunks of flesh and bone. Foxes and cats were known to discard wings and favor feet and bills. The identities of quail eaters could also be inferred from scat and regurgitated owl pellets. It was demanding and dirty work: sieving, washing, and sorting remains of meals after a long day in the cold. Human predators were a special problem: they took pains to leave no sign. However, poachers were often betrayed by gossip in a community where everyone knew everyone else's business and liked to talk about it. Errington's periodic visits to local taverns were not just for rest and recreation.[57]

Errington knew the general techniques of tracking and reading sign from his experience as a trapper. However, the particulars of census and mortality study were nothing that a professional trapper would need to know, and these had to be learned as science, as Stoddard had learned the diagnostics of broken eggs. Errington reckoned that Stoddard's help saved him a year's work of learning on the job.[58] Wildlife ecology thus combined the hunter-trapper's practical know-how with the ecologist's generic knowledge and precise empirical habits. Albert Gastrow's role in Errington's project at Prairie du Sac makes this point. Despite his experience of animals as a hunter and woodsman, there were things in the science that Gastrow could not do. In census work he was reliably precise—finding animals was a hunter's skill. But in mortality study his field records often fell short. Kill sites, for example, he tended to record as "usual remains"—a characterization good enough in hunting but useless for science.[59]

Forensic ecology was a kind of resident observing that went well beyond what local residents would ever need or care to do. Errington looked into every physical and biotic feature of study areas that might be relevant to encounters of predators and prey, combing the evidence "to extract the significant details of the ecological drama . . . taking place." It required "at times, an almost insatiable thoroughness."[60] He recorded details of local weather events and of the kinds and numbers of predators resident in the study area or passing through (e.g., flights of Cooper's hawks). He also estimated the abundance of predators' staple prey species—rabbits, mice, other small mammals—from which he could gauge how urgently predators would attack the more evasive quail and how likely they were to succeed. Exact and comprehensive knowledge of the total situations in which acts of predation occurred was vital to inferring causes from fragmentary evidence.[61] A tiny detail might to a knowing Sherlockian eye tell the true—the inside—story. Errington's ecology was both a situated and a situating science: observers and observed were cohabitants of a shared environment, and the total situation was as much a subject of the science as its resident quail.

Situation was no less vital to Errington's "threshold" theory of predation. The idea was simple. Quail residing in environments with forage and good cover nearby were essentially secure from attack: no predator however skilled could touch them. Birds residing in environments lacking adequate cover were, in contrast, doomed to be an easy meal for any predator that happened by. They were, as Errington put it, "spilt milk." And for every microhabitat there was a precise number of birds that could live securely in it, and that number was the "threshold of security from natural predation." It was not simply food supply that set limits to population densities, as in the familiar concept of "carrying capacity"—the term and concept were borrowed from range management.[62] Nor was it the density of predators, as many sportsmen believed. Security was a property of whole environments.

Traditional single-factor conceptions of predation were, as Errington put it "akin to grazing."[63] Encounters between predators and prey were like encounters of cows with tufts of grass. Grass, once spotted by a hungry cow, would be eaten—no defense, no escape. In this grazing model, how much was eaten depended only on the number of eaters. Particulars of situation, cover, and defensive strategies were irrelevant externalities. This model was obviously a fiction, but a useful fiction for those in whose interest it was to believe it, like advocates of game stocking and predator roundups.[64] The one animal to which the model did in fact apply is us—the "exterminator species" and "future eaters."[65] To all others a situated theory of predation applies. Prey species had through long cohabitation with predators evolved ways of using features of their environments to survive and reproduce. In environments that are reasonably intact, some will be eaten, but never all. It is the total environment that matters, and models that omit context—like the model of cows and grass—are misconceived. Errington in effect resituated an anthropocentric and decontextualized conception of predation in the complex environments in which dramas of life and death normally play out. His theory of thresholds of security was a theory of animal residence and, in the case of game animals, a theory of human coresidence as well.

Errington's conception did omit one feature of the total environment: all the contingencies of climate fluctuations, extreme weather events, and random disruptions. For Errington's purpose this was a reasonable omission. He was interested solely in the dynamics of predator-prey encounters, so deaths from bad weather and accidents were extraneous and had to be factored out of calculations of thresholds of security. This Errington achieved by adding back to his April census numbers the numbers of "emergency" deaths by causes other than active predation. These adjusted numbers—the "thresholds of security from ordinary predation" ("ordinary" was added to underline the adjustment)—were the numbers that had theoretical meaning.

Although the distinction between ordinary and emergency deaths was in principle straightforward, in practice it was anything but. Assigning individual deaths to one or the other category required total and firsthand knowledge of microweather conditions, the presence or absence of various animal species, and random occurrences in an area of almost nine square miles. Inevitably Errington's assignments were often a matter of his individual judgment. His lifelong experience of nature gave him confidence in his judgment; however, those who lacked that experience could only take his reasoning on trust. To achieve the scientific precision of numerical thresholds, Errington engaged in science at the very edge of subjectivity—or so it seemed to some, including Aldo Leopold, as we'll see.

Errington's confidence in his theory derived as well from the manner of its discovery. The idea of precise microenvironmental thresholds was a surprise, Errington recalled, and more or less imposed on him by his field data. His view of what limited population density was at first the orthodox one of "carrying capacity": food supply determined its upper bound; and below that maximum, populations of prey varied inversely with numbers of predators.[66] Field data from the first three years at Prairie du Sac bore out the prevailing "grazing" view. It was the results of the 1932–33 season that first caused Errington to question the common wisdom: fall and spring census numbers were virtually the same as the preceding year, yet predator populations were dramatically higher. Quail populations should have declined accordingly, yet did not.

Errington was also surprised to observe that it was not just the overall number of spring survivors that remained constant, but the number in each one of the eight covey territories. It appeared that each particular patch of habitat afforded secure residence for a definite number of birds and no more.[67] Field data from the next two winters seemed to confirm that stable thresholds were not a fluke but a real property of environments. Errington was delighted with his unexpected discovery, though also a bit shaken by seeing settled scientific theory overturned. Hunters, he knew all too well, could believe unshakably in impossible notions—but scientists? "I'll confess," he wrote Stoddard, "that a number of my own views suffered reversals when I got the data spread out for analysis."[68] It felt like a major discovery about the ecology of predation.

However, these first years proved to be a "period of comfortable generalizations."[69] From the mid-1930s, threshold numbers jumped abruptly from one level to another, from a seemingly stable low in 1935–37 to an intermediate plateau in the late 1930s, and then fluctuated wildly from one year to the next, with no parallel changes in the physical or biotic environment that Errington could see.[70] He did not waver in his belief that precise thresholds of security were a property of environments: but what exactly was "environment"? Was it something that quail could perceive and humans could not? Or was it something in the birds themselves—some undiscovered feature of animal social psychology? Changes in the birds' tolerance of intraspecies crowding seemed a likely possibility, or changes in "traditions" of self-defense that were, or were not, transmitted from one generation to the next by experienced older birds. Perhaps "environment" could not be ecologically defined at all but was simply whatever bobwhites made of it. Errington leaned toward a biobehavioral explanation—something in the birds—but remained unsure.[71] He nonetheless never lost faith in the idea of exact thresholds of security, even as more data only deepened the mystery.

Linking a quantitative theory of population to an empirical method that relied on personal know-how and discretion was a bold and risky move, but it is understandable in the context of Errington's dual life history. As a trained zoologist he would of course value numerical precision and general laws. But from his early life as a subsistence trapper, he had learned to trust the personal experience and judgment of contexts on which his own survival had depended. Errington always felt that his chief advantage over completely school-reared scientists was his trapper's lived experience with wildlife. "In general," he wrote, "I feel that I am distinctly at my best doing intensive research on a local scale, contrasted with research of the state-wide survey type as done so well by Leopold."[72] He was a proper scientist, yet felt most secure and confident when he was doing a hunting kind of science. Stoddard likewise believed that his confidence in taking controversial positions on controlled burning and predator control derived less from his knowledge of scientific method than from his long experience as an outdoorsman-naturalist.[73] Because their formative experiences of animals in context preceded formal training, they could draw equally on either one, or both together. A deeper look into Errington's life history is in order.

ERRINGTON: A LIFE HISTORY

Errington came to advanced study in zoology by a winding path. He was the scion of immigrant Swedish homesteaders on both sides of his family (the Erringtons had been Ericksons in Sweden). He grew up in the college town of Brookings, South Dakota, with his mother and stepfather, an Irish-American who made ends meet as the proprietor of a small ice cream parlor.[74] Though Paul lived in town, he spent as many days as he could at his maternal grandparents' old homestead farmhouse at the edge of town on Lake Tetonkaha, a place that was easily reached and full of things and creatures fascinating to an exceptionally curious, intelligent, and observant youth.[75] The wet prairies of Minnesota and the Dakotas were at the time a lightly settled area of what fifty years before had been a sprawling wild of marshes—the vestiges of vast glacial lakes—and glacial pothole ponds that constituted one of North America's great flyways and breeding grounds for waterfowl. A "duck factory," some called it. Though greatly diminished by large-scale reclamation, prairie marshes remained islands of wild in an enveloping agricultural landscape, and an important part of the region's human ecology. Country began at the edges of small towns, and hunting and outdoor recreation were central to community life and a rich cultural soil for wildlife science. The marshes were a factory not just of ducks, but also of sportsmen-naturalists and ecologists. This inner frontier, like central Florida's, was a fleeting moment in the continent's environmental history, and Errington's and Stoddard's generation was about the last to experience it. Marshes and their animal inhabitants were Errington's lifelong love.[76]

When Paul was eleven, his stepfather gave him a .22-caliber rifle, and he began to spend weekends wandering about "plinking" in the countryside. This was a common boyish amusement, but for Errington it had a deeper import. When he was eight, a crippling case of polio left him with a withered right calf and ankle. After a year he could get around, but with difficulty, and the gift of a rifle was an incentive for him to be more active in his recovery. By the age of fifteen Paul could outwalk most of his friends and was hunting and trapping in a cruising range of ten and occasionally thirty miles from home. Outdoor life and hunting were for him not just delightful recreations but the gift of a fully active life that he would otherwise not have enjoyed. Though he lived in town, the world of ponds, marshes, and prairie streams was his true backyard.

Paul also discovered that he had a knack for hunting. The conservationist and political cartoonist J. N. "Ding" Darling recalled how, hunting in the Dakota marshes, he would hear "now and then of a boy named Errington who could shoot like nobody's business."[77] Paul began trapping and selling pelts, he recalled, because it was an "acceptable excuse for spending time in the out-of-doors."[78] It also put cash in his pocket and food on the family's table; however, the deepest satisfaction came from the visceral pleasures of being out in nature and part of it all. When Paul was twelve, he was granted permission to use the family shotgun to shoot prairie chicken for family dinners. Since he had no idea what the animal looked like—a fact he had neglected to mention and could not confess without losing face—he brought nothing home. So when a farmer he was helping pointed out a group of the birds, young Paul looked long and hard. It was, he recalled, his first wildlife observing.[79] "Kid Hunter" was no naturalist: he knew nothing of science and was not a willing scholar. School was the irksome obligation that kept him from the places and pursuits he truly loved. Yet in his hunting and trapping, he developed habits of observing and reasoning that would also be those of a wildlife ecologist.

Errington's experience of nature was not unmediated. From his earliest years he had been an avid reader of popular romances of "wilderness" adventure by Jack London and especially Ernest Thompson Seton. He acquired a romantic taste for flirting with danger, to prove that he had what it took to survive—nothing foolhardy, but close to the edge. He liked to imagine areas of marsh or brushy moraine as "wolfish wilderness—deliciously chilling."[80] It was an imagined life in a natural world that was exciting and dangerous yet never ended in tragedy. So at the age of fifteen Paul determined to live as a trapper-naturalist and, screwing up his courage, announced to his mother that he intended to drop out of high school to do full-time what he loved and did best. Spirited arguments ensued, which ended with Paul reluctantly agreeing to finish school first. Meanwhile, he prepared himself for a life in the wild. He spent a cold December alone on an island in Lake Tetonkaha, trapping and eating what he could kill, and struggling to stay warm and survive. (He was near enough to home to be rescued if need be.)[81] On graduating from high school in 1920, he spent seven winter months in the Big Bog region of northern Minnesota and, to his mother's professed surprise, survived the experience.[82] He was by then financially on his own, living by trapping the marshes and river bottoms of his Dakota big backyard.

At last in 1921 he realized his dream of going to the boreal wilderness of northern Canada, to scout a good hunting ground in which to live and work as a professional trapper-naturalist. That adventure he survived as well—just barely, as we will see—and returned, chastened, to South Dakota and the tamer but more secure life of farming and trapping.[83] Though Errington had been disabused of his romance of residing in the great northern "wilderness," he still meant to somehow live an outdoor life: if not as a trapper-naturalist, then as a professional naturalist. Gradually, however, he realized that experience of animals in nature was not a sufficient basis for a professional career; he would also need formal education.[84] So at the age of twenty-one he became a part-time student at South Dakota State University, commuting from the family farm and paying his tuition fees by trapping and selling pelts.

The transformation of Kid Hunter into a credentialed zoologist was not easy. His ingrained conviction that schooling was the enemy of outdoor life made it hard for him to see how the two were in fact compatible. He disliked classrooms and had to be coaxed to stay there. His adviser, the ecologist Earl O'Rourke, kept Errington taking courses in biology by slyly pointing out that what he learned in them would make him a better hunter-naturalist. He was also encouraged by a maternal uncle, Aaron Johnson, who was a plant pathologist in the US Department of Agriculture. It was Johnson who urged Errington, in 1929, to apply for the SAAMI fellowship at the University of Wisconsin, where Johnson had once been a professor.[85]

There were moments of backsliding, as when, trapping in the Cheyenne River breaks in western South Dakota, he thought how splendid it would be just to quit college and take up residence among friendly neighbors in a semi-wild area of great beauty and farm, trap, and naturalize. But there were countervailing pleasures on the academic side, as when he got a scientific collecting permit to get specimens for the college museum, and his "old meat-getter shotgun acquired the respectability of a scientific instrument."[86] He discovered that science made an outdoor life socially respectable, as commercial trapping had in his youth. The role of outdoorsman-biologist was a viable one, as the role of hunter-naturalist no longer was. School was no longer a detour but a path. Thus did his life's winding road take him to university and to the bobwhite project.

One particular episode was a turning point. It occurred during his sojourn in the Canadian "wilderness." As Errington recalled the events, he was preparing to canoe down the Kenogami River to a promising hunting area and was glad to join up with two Indians who knew the country and were going that way. But he kept falling behind on the portages (it took him two trips to do what his companions did in one), so at one rest stop he determined to get a head start. The older of the Indians, who was friendly and spoke English, instructed him to run the first two rapids he would encounter, then portage around the next. Errington ran what he took to be the first, then paused at the head of the next, wondering how to proceed. He could see no good channel but assumed that the sound of roaring water ahead was the third rapids farther downstream. "I was sitting in the canoe trying to make up my mind where to run that rapid when the Indians came packing along the shore. The old man asked me sharply what I was doing there. I replied that I was looking; I got out and started portaging." What lay just ahead, he soon saw, was a long series of rapids and falls in a narrow rocky chasm from which no man could escape with his life. The sight of it from the portage path left his mind "benumbed." It took some weeks to admit the full truth: he did not have the skill and experience to live as a trapper in such a genuinely dangerous environment. That admission was, as he put it, his "psychological metamorphosis."[87]

Errington's close encounter with extinction revealed to him that the romantic literary images of "wilderness" as a place of adventure and personal testing were a dangerous fantasy. "The Canadian wilderness had neither an inimical nor a friendly aspect—only an impersonality and an unfeelingness that I often compared with interstellar space. The northern wilderness offered no personal challenge to anyone. . . . If anything was there that could live, it lived; if it could not live, it did not. . . . I knew where I would stand in that wilderness. I would live or die on the same terms as the Indians, the wolves, the hares, the grouse, the chickadees, and the rest—living if I could and dying if I had to."[88] With that new self-knowledge Errington gave up the romance of living as a hunter-naturalist in the northern "wilderness" and returned to reside in his own working landscape, "grow up some more," and pursue further schooling in science.

What Errington learned as a trapper-naturalist informed all his subsequent work as a wildlife ecologist: his fascination with individual encounters of predator and prey, and unwillingness to intervene in dramas of life and death; his devotion to close observing in situ; and his insistence on the importance of total environment in animal life. He was fascinated with individual acts of predation as a scientist because his own life as a subsistence trapper had consisted in such acts. He understood the life-and-death importance of cover and context from personal experience of how skillfully animals used their knowledge of home environments to evade predators like himself. He did not intervene in dramas of predation because such acts supposed a human-centered view of nature that his own near miss with death on the Kenogami River had shown to be false. It was not indifference to life and death but respect: knowing he was not a master of nature but part of it and bound by its rules. Life experiences thus became elements of his science, as Stoddard's experiences had become part of his. Aldo Leopold is another such case.

LEOPOLD: LIFE AND LAND HEALTH

Although Leopold was of a mind with his two younger colleagues in their love of hunting and outdoor life, and dedication to conservation, he was never drawn to the kind of intensive situated observing to which Stoddard and Errington were devoted. He liked to operate on larger issues and on a larger stage, as in his regional game survey and his game treatise and land-health essays. No resident observer, Leopold was nonetheless a deeply residential thinker. His philosophy of land health and stewardship was in effect a theory and program of coresidence of humans and animals, in which every species, including predators, had an equal claim to belong. It was thus of a piece with Stoddard's practices of managing land for game, and Errington's environmental ideas of cover and security. Stoddard's human ecology of cohabitation and Errington's ecology of predators and prey were the foundation Leopold's environmental philosophy.

The first principle of Leopold's land ethic was that a healthy and livable place is one that is ecologically intact. All species are legitimate and desirable residents, including predators, because they had always been there and were essential to the long-term well-being of all species, including our own. Having invented agriculture, humans cannot help but change the places we inhabit, Leopold acknowledged. However, stripping environments of all but the species useful to us by monocropping or killing off competing predators only leaves them less fit for human residence, and that *can* be helped. "The land is a factory," he wrote, "but it is also a place to live, and wildlife helps make it a good place."[89] Wild animals of all sorts were "nice to have around." Residence was thus at the core of Leopold's environmental philosophy: how species cohabit places, and how our own species should live and let live.

The second key principle of Leopold's land ethic was likewise deeply residential: responsibility for maintaining healthy environments rests with human residents, farmers especially. Resident farmers made the best conservationists, Leopold argued, because they knew the land they worked more intimately than visiting experts ever could. And as landowners with the legal rights of freehold, they alone could choose to make it hospitable to wild species like quail, owls, muskrats, and mink. They alone could choose not to plow fields fence to fence and to preserve brushy edges and wet spots as cover for wild creatures. Such untidy practices were usually taken to be signs of a farmer's laziness or incompetence, but to Leopold they signified that a farmer grasped the principles of land health. "Localized scientific reasoning" he called it.[90] *Residential* scientific reasoning would be an equally apt term. Farmers were not ecologists and did not need to be; but as observant residents they were best able to see, in Errington's words, what made the wheels of life go round. The Leopold family's efforts to restore to health a worn-out sandy farm were Aldo's small-scale attempt to practice what he preached.[91]

As Errington's science grew out of his life experience, so did Leopold's ethic of land health grow out of his.[92] Leopold, too, was born and raised in the mosaic landscape of the upper Midwest, in a German-American professional family in the rural town of Burlington, Iowa. Corn Belt Iowa was a more settled, tamer place than the Dakotas; but nature and wildlife were accessible, and no less a vital part of local culture. Aldo developed an early and passionate attachment to outdoor life, and as an avid sport hunter and outdoorsman, he determined early on to pursue a livelihood that would be an outdoor life. Leopold's occupational horizon was wider than Errington's—the difference between their middle- and lower-middle-class locations—and he decided to pursue the profession of forestry. So whereas Errington headed north to Canada to become a trapper-naturalist, Leopold went east to Yale College and the Yale School of Forestry, and from there to the Southwest as a fledgling forester with the US Forest Service. There he was put in charge of managing private livestock grazing in that region's vast national forests.

The mountain forests of New Mexico were yet another inner frontier: a large island of public land that was wild yet accessible and used. Overused, in fact: Leopold's chief problem was putting a brake on the destructive overgrazing that was the custom among the region's Hispano and Anglo ranchers. Against determined opposition he put into practice the progressive ideals of sustainable resource use that he had learned at Yale, ideals exemplified by the writings and work of the forester Gifford Pinchot.[93] He engaged in predator control (coyotes, wolves) as was the custom, and strove to balance the competing claims of economic production and environmental health. Gradually, however, he became disenchanted with the conflicts and compromises of managed use and was drawn to a preservationist ethos, of which the naturalist and writer John Muir was the leading voice. Leopold grew less sure that predator control was a good thing and began to see environments as places for all species to live in, rather than as places for growing favored species for human harvest. In New Mexico he sought to establish forest preserves closed to livestock grazing and strictly for recreational and aesthetic use. These activities, predictably, did not improve his relations with local ranchers, or for that matter with his Forest Service colleagues. So in 1924 he returned to his native region, as head of the Forest Service's Forest Products Lab at Madison. Unsurprisingly, desk work proved not to his liking—it's unclear why he ever took the job—and in 1928 he resigned from the Forest Service to become a consulting game expert and to oversee the SAAMI projects.

It was long the default view that Leopold's land ethic derived from his experiences in the mountain forests of the Southwest. Deer, wolves, "the fierce green fire" in the eyes of the dying wolf—it's an oft-told tale. It seems quite clear, however, that Leopold's ideas of human and animal ecology were the operating principles not of contested wilds but of the settled human landscapes—the "second nature"—of his home region. Baird Callicott and Eric Freyfogle have made this point so clearly and forcefully that I will simply quote them:

> In the Southwest, the internecine conservation battle lines were drawn between efficient resource exploitation and wilderness preservation. But the Midwest to which Leopold returned in the 1920s lacked extensive public lands and national park grandeur. The great pineries of Michigan and Wisconsin had long been felled. The great prairies of Illinois and Iowa had long been plowed and planted to annual grains. . . . [L]ittle wilderness was left to preserve. The landscape had been largely carved up into privately owned smallholds. How could a conservationist apply either the Pinchot paradigm or the Muir paradigm in this kind of country? . . . Big wilderness was out of the question; so was big-scale forestry. What did seem possible, Leopold decided, was the integration of a degree of wildness into the working landscape mosaic of cultivated fields, pastures, woodlots, and wetlands.

This "third conservation paradigm, this middle ground between unsullied wilderness and unrestrained exploitation," Leopold would in time call, simply, "land-health."[94]

That seems to me exactly and self-evidently right. Leopold's philosophy of human and animal residence, like Errington's residential concept of thresholds of security, was a vision formed in the experience of growing up and living in a working landscape of small fields, woodlots, and glacial ponds and marshes—a "second nature." It is no accident that the upper Midwest, with its checkerboard of wild areas and college towns, was a cultural hearth of wildlife studies and of ecology generally in the United States. Stoddard, Errington, and Leopold were uncommon exemplars of a common social type in that mosaic landscape. And the science they espoused, and in their different ways practiced, was likewise a mosaic of cosmopolitan learning and situated lived experience.

Errington's wildlife science and Leopold's land ethic developed together in their collaboration on the bobwhite project. Stoddard's and Errington's fieldwork provided the empirical substance of Leopold's general concepts—seventy-seven and thirty-nine citations, respectively, in his *Game Management*, more by far than any other source.[95] Leopold was quick to see the larger implications of Errington's evidence that it was the total environment, and especially cover, that enabled game birds to flourish. It was "the most important recent discovery in game management," he advised the editor of *American Forestry*, "not an easy idea to convey to the public, but an exceedingly important one."[96] In a letter to Errington from 1933, Leopold hints of what his developing idea of farmers as conservationists may have owed to Errington's science:

> We habitually approach game research in terms of individual species studies. Why use this unit? . . . You are trying to reach farmers. Why not, therefore, use land units, and treat of all species thereon. This comes much nearer being applied ecology. For instance, it is thinkable that you might select certain farms, and on them carry out, as an ecological whole, a study embracing all the species topics you now have in hand. . . . Thus instead of quail census, fox food, dove abundance, pheasant kill-rates, etc. all being scattered in different places, they would all be studied where their total, as well as their separate values, would yield something new. . . . Politically it would enable you to talk more about average farms, and less about special interests such as sportsmen's groups.[97]

Although the two saw eye to eye on the principles of land health, Errington resisted all Leopold's exhortations to extend his science to practical field demonstrations. He detested hunting politics and did his best to steer clear of it. "I don't have a nervous system that can endure militant ignorance and the punishment that goes with seeing truth abused," he confessed to one correspondent. "I simply can't 'take it' and sleep afterwards or concentrate . . . on . . . my own work."[98] He adopted the ecological concepts of Leopold's philosophy, however, even if he did not put them into practice. Replying to a sportsman wanting to know how to ensure abundant game birds on his property, Errington wrote in distinctly Leopoldian terms: "[K]eep the environment fit," he advised, and apply "simple management principles dove-tailed in with agricultural and community practices."[99] Nothing in Errington's dual life as subsistence trapper and ecologist inclined him to engage personally in practical interventions, while everything in Leopold's life as sport hunter and ex-forester did. On healthy environments they saw eye to eye.

The two men likewise held somewhat differing views of wildlife science. Leopold looked to biological disciplines like population biology and ethology for objective methodologies, and valued general laws and theories. Errington, in contrast, favored intensive observing and reasoning directly from facts, and distrusted deductions from theories that he had to take at second hand, on trust. So he declined to take Leopold's advice to expand his investigation from winter mortality and predation to summer mortality and an all-season population dynamics. He declined as well to make use of the theory-heavy science of population biology, which was expanding dramatically in the 1930s and, Leopold believed, making Errington's science out-of-date. Leopold stated publicly at a conference in 1947 that the "old" methods of census and mortality study could never give a complete picture of predation, and ecologists must adopt the "new" methods of population biologists, such as analysis of age and sex cohorts. The "failures" of the 1930s were leading imperceptibly to improved field practices.[100] He did not name names, but must have had Errington in mind. In fact, Errington had made an effort to engage with the large literature on population cycles of animals in nature, but he found it impenetrable and unhelpful. A "scientific shadowland," he called it, of abstract mathematical models based on secondhand data from published and archived sources (e.g., fur-trade records) that lacked the authority of personal experience.[101] It's no surprise that Errington the ex-trapper would favor a Sherlockian practice of clues and conjectures, while Leopold the professional land manager would be attached to Galilean generalities.

Errington's collaboration with Leopold frayed in the early 1940s as the two labored to write the final and comprehensive report on the Wisconsin investigation. The chief bone of contention was Errington's practice of distinguishing "emergency" deaths from those by "ordinary" predation. In Leopold's view Errington was drawing unwarranted conclusions from imperfect data and allowing hypotheses to "sneak in" as facts, trusting uncritically in individual experience and judgment. Errington, in contrast, thought Leopold's reluctance to accept his reading of the facts stemmed from his imperfect mastery of the field data and his lack of personal experience of how they were gathered.[102] Leopold, the trained professional, valued facts and generalizations produced by methods that by common agreement among peers were legitimate, and distrusted claims based on individuals' personal qualities.[103] Errington, the man of woods and marshes, valued the authority of personal experience and was untroubled by generalizing from imperfect data so long as these were produced firsthand by experienced observers. After a long back-and-forth correspondence, Leopold finally withdrew from their joint project, leaving Errington (reluctantly) to publish under his name alone.[104]

CONCLUDING THOUGHTS

The case of bobwhite ecology is both like and unlike the other cases in this book. The engaged relationship between observer and observed that has traditionally defined participant or resident observing is absent: no habituation or mimic quail acts, no individually named birds, no little stories of quail nest and covey life. Yet it is a case of resident science. Stoddard's and Errington's field practices were both situated and situating: quail were studied in their natural environments, by observers who were either resident or as good as. Stoddard's strategies of land management, Errington's theory of cover and security, and Leopold's land-health ethic were all about coresidence. However, issues of coresidence in this case arise less with the avian than with the human residents of study areas. Small farmers in Wisconsin and estate owners and tenant farmers in Georgia were active participants both in quail ecology (as curators and hunters of birds, and as creators or wreckers of their habitat), and in the science, as subjects and co-investigators.

The residential ecology of bobwhites was more complicated than any we've met in previous cases. Kiriwina, Chicago, and Cornerville were human ecologies; and Gombe (despite the ambiguous banana tree) was an ecology of nature. The sites of the quail projects were both. Bobwhites were a species somewhere between wild and commensal: wary and evasive, yet favoring the disturbed habitats of farm and woodland. Wildlife ecology was in this case (though not in all cases) an ecology of cohabitation. Whereas the banana tree briefly disarranged the concept of residence in Gombe, bobwhites' farm-loving habit more lastingly expands the meaning of residence not just in the Red Hills and Driftless Area, but in all the varied landscapes of humanized "second nature"—which in the contemporary world means just about everywhere. We are thus led to a larger and more inclusive conception of residence in science.

The case of wildlife science also elaborates the idea that practices of life and science can overlap. We have here not one but three life histories, generally alike but differing in their scientific and professional practices, as well as in the timing and balance of their formal education and worldly apprenticeship. Stoddard, Errington, and Leopold as lifelong outdoorsmen and environmentalists had much the same experience of the natural world and the same views of human and animal cohabitation. Yet the three differed in when and how worldly apprenticeship and formal learning entered into their life histories, and these differences led them to pursue different occupational paths: Stoddard to managing game-bird habitats, Errington to the ecology of predation, and Leopold to an environmental philosophy of human and animal coresidence. Their careers in wildlife studies constitute a natural experiment in how life histories and experience become modes of science.

The experiment is in fact almost too good to be true. Stoddard's "education" in nature (as he would put it) was of the three the most extended and formative. His formal schooling ended early and left little residue, and his halting entrée into professional work as a museum taxidermist was ultimately a lesson that scientific ornithology was not for him. What he did as a professional wildlife manager was consequently virtually a continuation of the apprenticeships of his early life. Errington's experience as a subsistence trapper was also early and deeply formative, but it drew him into, not away from, formal education. Although he came belatedly and reluctantly to advanced study in science, he entered quickly and pleasantly into a scientific career, assimilating the values of professional science—empiricism, quantification, cause-and-effect reasoning. Yet as a wildlife ecologist he remained always most comfortable with a subject and with empirical practices that drew from his experience as a trapper-naturalist. Leopold's formal and professional schooling, in contrast, was extended and formative, imbuing him with an abiding trust in the values and methods of professional science, which survived his disenchantment with "scientific" forestry and shaped his metamorphosis into an environmental philosopher and activist. He remained largely what he had been schooled to be: a theorist and organizer. Each man was thus drawn to the scientific or professional practice that was most congruent with his particular mix of worldly apprenticeship and formal schooling.

These patterns of association between life history and science, should they hold up to further study, support and further enrich the patterns of life history and science in the cases of Malinowski, Anderson, Whyte, and Goodall. In each, practical experience and formal schooling were formative in their choices and practices of science, varying with the particulars of their life histories, subjects, and situations of resident observing. These connections between practices of life and those of science do not prove cause and effect, of course, but remain conjectures from limited evidence. But conjectures will do for now. Whether a significant connection with life experience is generally a feature of deeply situated and resident science—or indeed of other less situated sciences—are questions for wider empirical study than my exploratory gambits here.

EPILOGUE: INSIDE SCIENCE

I opened this book with the premise that context always matters: everything that happens in nature and in human society depends on context and is therefore best studied and understood scientifically in context, as it is in life. I will close the book with a brief overview of what its cases of resident observing tell us about the particulars of contexts: not just *that* they matter—a fact of daily experience—but *how* they do. What constitutes context in any instance varies a good deal, but in every instance context shapes the course of actions and events, often decisively. Contexts are not just the places in which things happen but are constitutive of what happens, and the nature of their agency is revealed by observing actions and events in place, from the inside out. If there is one take-home point in my stories of resident observers, it is that.

The *how* of context is most generally and concretely set forth—conceptually organized and exemplified in vivid stories from life—in Malinowski's account of Trobriand law. (His slender *Crime and Custom in Savage Society* is for that reason my favorite of all his monographs.[1]) Most striking, I think, is the stark contrast between the ideals of law and the realities of its practice, as readers will no doubt recall. Officially, codes of conduct were laws inflexibly applied (and in principle obeyed to the letter) irrespective of context. That was the view that visiting ethnologists took from interviews with official informants. The inside view of resident ethnographers, however, was that law in practice was often bent to unruly human passions and messy situations of communal life. The accommodation of the matrilineal law of "Mother Right" to the human realities of "Father Love" was paradigmatic. By law, rights to property and political and sacred authority were inherited strictly in maternal lines, from husbands to maternal nephews. But because in life husbands were often emotionally closer to sons and favored them over legal heirs, breaches of the law were common and unavoidable. Custom permitted such breaches to be overlooked, so long as the cosmic order of law was not overtly challenged (in which case it applied without recourse). Universal principles were thus allowably bent to the particular contexts of family and social relationship.

Similarly with the law of exogamy, which forbade sexual relations between first cousins—siblings by law—though it seems to have happened often enough. Here again breaches of law were by custom overlooked, if couples acted discreetly. In this case, however, accommodation to circumstance was not just the custom, but was embodied in practices of ritual magic—a kind of common law—which if exactingly performed could undo the evil consequences of incest and thus allow for lenience in individual situations. (Still other ritual magic was available to undo the effects of the first.) In an official world of absolutes, a system of flexible accommodation allowed for the intractable complexities of human contexts—a system whose complexities mirrored those of life. It was a system that was opaque to outsiders and that opened only, and then slowly, to persistent resident observers.

The power of context was similarly revealed to resident observers of modern societies. William Whyte's street-corner society displayed the same conflict between official principle and pragmatic accommodation to human contexts. Group members steadfastly denied that they had a social hierarchy, even as leaders openly manipulated contexts of group activities to shape individuals' behaviors and social standing in the group—as in the jeering and cheering that accompanied competitive sporting events. Self-confidence, leadership, and physical skill were thus not just innate qualities but depended as well on the social contexts in which they were exercised. The power of situation was most clearly and dramatically demonstrated in Bill Whyte and Doc's experiments using calculated manipulation of individuals' social interactions to restore their mental well-being and ability to perform. This particular power of context was a general truth revealed by inside study.

In animal life, context is a mix of natural and social environment. Paul Errington showed how physical context—cover—was decisive for survival of bobwhite quail: not the number and frequency of encounters with predators (as was commonly believed), but the particular contexts in which encounters occurred. Birds with access to good cover were virtually invulnerable to attack, however many predators were on the prowl. Those without cover were doomed to be a meal for any passing raptor. Errington's theory of thresholds of security was thus a theory of context. And when variations in threshold numbers failed to correlate with changes in physical environment, Errington expanded his conception of cover to include the psychosocial aspects of covey life: in particular, birds' tolerance or intolerance of intraspecies crowding. Context—cover—was what resident birds themselves made of it. (That at least was the idea; proof remained elusive.)

Context was no less a decisive element in the actions and events of chimp society, but in that case it was mainly social context. Stock behaviors of greeting and deference in social encounters varied with who was present—males or females, family members, rivals, allies. Such variations at first seemed just typically fluid, but the more complete knowledge of resident observing revealed the existence of a community of the whole, in which every animal knew its position with respect to every other and adapted its behavior to the particular contexts of the small groups in which chimps lived. It was the community as a whole that was the operative context of life in Chimpland, and that was an insight obtainable only by being continually present in the contexts of chimp everyday life, observing foraging groups interacting in all their permutations and combinations.

Context is obviously an almost infinitely varied factor in animal and human life, different with different actors—animal species and ecologies, human societies and ways of life—and different kinds of action. But context of whatever sort in all its variety is always a factor, and often a decisive one, in what there is and what goes on in nature and human societies. And that "how" of context is best seen and understood by observers who together with their subjects are in place where actions are initiated and play out. It is true that desituating practices of science are powerful and essential, especially at the end of investigations, to connect more precisely single-factor causes with effects. But in the world of actual events, causes and effects are seldom precise or single, but are modulated, amplified, or deflected by circumstance; and accounts that omit situation are almost certain to be incomplete.

I would thus be remiss (and inconsistent) if I did not put this history of resident observing into its larger, historiographical, context. It is clear enough, I think, that the practices of resident observing occupy a small, if significant, corner of a far bigger and more diverse world of practices that bring observers and observed together in situ in some way short of actual coresidence. For the purposes of this book, I chose cases in which observers were materially cohabiting with their subjects. I wanted cases that were well defined and exemplary, yet spoke to issues of context and situating more generally. So I stayed away from all the halfway means of situating that I briefly surveyed in chapter 1 to highlight the differences between those and fully resident practices. In a larger exploration of context and situating in science, modes of partial or even virtual presence may be quite as revealing as fully residential ones. So the question now, at the end of this initial foray, is how to get from the special topic of resident science to the large and sprawling one of practices that operate in some way or other inside the objects of observers' study. "Inside Science" might be a useful label for that extended type and domain of practice. I've used "inside" observing casually where a more permissive term than "resident" seemed right, but the general idea calls out to be developed in a more systematic way. The omnibus category in effect resituates resident observing in the more diverse world of more or less situated practices from which, for the purpose of this book, I plucked it out. The question, then, is how do we accomplish that resituating.

One might begin by reconsidering the practices that figured in this book as incomplete or transitional—intensive surveys (social, ethnological, ecological), community studies, expeditions, seasonal sojourns of fieldwork. Although my point was that these recurring practices were less than resident observing, there are likely also ways in which they were more, as in their periodic alternations between local and cosmopolitan locales. Participant observers sometimes combined periods with their subjects and periods at home to shuffle data, think, and write, or just refresh (Malinowski in Australia, Goodall in Cambridge). As I noted in Goodall's case, it is not just contexts in the field that matter for the science, but also those at home. How exactly each matters will depend on how the going out and the coming back are managed and experienced (for Malinowski home was mostly a welcome respite, for Goodall mostly a stressful duty). We thus put in motion the concept of residing.

Another way to expand the concept of residing would be to look at studies of animal species less tolerant of human presence than, say, quail, or that are by nature or human design our settled cohabitants. Ethologies of domesticated, companion, and commensal species are obvious possibilities—the weedy cohabitants with our own weedy species.[2] Konrad Lorenz's studies of greylag geese unnaturally imprinted on himself are a famous case of barnyard science. And the researches of Althea Sherman on chimney swifts using a home-built swift tower and those of Margaret Morse Nice on song sparrows were in their time models of science in backyards.[3] Backyard and barnyard science were once the common practice of natural historians who aspired to a cosmopolitan scientific life but were obliged by their rural occupations of clergyman, teacher, farmer, or landed gentry to improvise a home science. Charles Darwin at Down House was famously a backyard family naturalist. And Sherman, Nice, and other homebound naturalists likewise turned the restrictions of spousal obligation to scientific ends. I elected not to include studies of backyard and barnyard animals in this book: residence in these cases seemed too literal, and backyards and barnyards were more open-air labs than "field." But in the more spacious domain of inside science, ambiguous spaces that are neither field nor lab are appealing subjects for historical study.

Limits of subjects' size and geographical scale could also be explored and probed. Resident observing would seem from the present cases to be a practice of the mesoscale. But are there ways in which observers could be said to co-inhabit with the invertebrates and small vertebrates of meadow turf and forest leaf litter? Not literally, of course: but local naturalists' minute and close attention to the inconspicuous activities of woods and fields puts them more than just metaphorically inside their subjects and yields results very like those of actual resident observing. Such close attention to small creatures and small places need not be done at home. Peter and Rosemary Grant's decades-long ecological investigation of microevolution in the finches—Darwin's finches—of the Galapagan desert islet of Daphne Major is a striking case.[4] The place is for humans utterly and forever uninhabitable: none of the necessities of life are there or ever can be. Yet in their exact and comprehensive measuring of slow ecological and morphological change, the Grants were unmistakably inside their subjects' unforgiving little world. Another case (and a personal favorite) is the ecologist Charles Elton's day-in/day-out, decades-long ecological survey of the tangled trophic and ecological relationship among the small creatures of Wytham Wood, Oxford University's nature preserve.[5] Elton and his team of observers did not live in the place, but in their total knowledge of Wytham's small inhabitants and their lives, they were inside in every sense but the one of literal residence. These subjects are just the few that I have happened to encounter in my wayward foraging. Systematic survey would certainly reveal more and doubtless better ones. Much of animal and human ecology is science of small worlds (as is much of science history).

On the other extreme of size and scale, we might ask if inside observing can be extended to migratory ungulate herds (wildebeest, caribou), far-wandering ocean creatures (eels, whales), or large predators (lions, wolves, bears). One man who thought he could live among Alaskan grizzlies ended up as lunch for one of them.[6] But Adolph Murie found ways of observing closely yet safely the lives of Alaskan wolves.[7] And advanced techniques of remote tracking of individuals and groups, though very different experientially from in-person walking and watching, can have results that are much the same.[8] Technologies of large-scale, even planetary, surveillance may be changing the meaning of being inside a place in even more fundamental ways. It seems clear enough that resident observing is not invariably a science of the mesoscale. There are subjects for historians at the extremes.

One could also enlarge the domain of situating science by pushing the limits of time. Most cases of resident observing, and all the ones in this book, are carried out in the here and now. But does it have to be so? Could inside observing be taken back to the there and then of the past? Historians like to say that the past is a foreign country.[9] And if that is so, would it not follow that the past is also a place that historians who know it deeply could be said to inhabit and observe? The historical sciences of archaeology, paleoecology, paleoanthropology, and evolutionary biology could be major provinces of inside science. Martin Rudwick has exposed the deep and rich history of purely imagined visits to scenes of extinct creatures living in vanished landscapes, with only fossil bones and plants to go on.[10] In the past seventy years or so, however, methods of excavating, collecting, and data analysis have been devised to give empirical substance to the ecological and social contexts of what became fossils and artifacts—contexts that were once overlooked and unthinkingly destroyed in the process of digging and collecting. Palynology—pollen analysis—and the spatial distribution of humble fragments of bones and artifacts across large sites are just two of the techniques that have transformed archaeology from a taxonomy of things into a science of natural and human ecology. And an array of isotopes starting with carbon-14 has taken the sciences of our planet's physical environment and biosphere—the context of all living things—back into very deep time.[11]

But what exactly (or inexactly) does it mean to say that archaeologists and paleoecologists "inhabit" the foreign countries of the past? Is it just a figure of speech? Or is there a sense in which it is functionally true? Not physically, of course, but the physicality of inhabiting is only half of it. The other and perhaps more important half is inhabitants' ecological and social relationships; and with sufficient empirical knowledge of these, investigators can be virtually in the situations of the past, feeling themselves in the middle of things and finding their way around, much as they would in their own present worlds. Such accounts of the past may always be in error, of course, but so long as they are true to what is known empirically, they cannot be dismissed on principle as fictive impositions. Badgers need not be physically present to make observers think of vicars. Charles Elton's analogical reasoning in animal and human ecology applies as well to places vanished and extinct. The past may be a foreign country, but it is not a different planet.

And what about the line between sentient and insensate subjects? I limited myself in this book to cases from the human and animal sciences, to which the concept of residing unambiguously applies. But might the more capacious concept of being inside extend as well to insensate material things? Could organic chemists, for example, be said to get inside and inhabit the complex molecules of plant and animal bodies? I can relate one such story from my own experience. Listening (as a graduate student) to the organic chemist Robert Burns Woodward recount his feats of total syntheses of such complex molecules as chlorophyll or vitamin B_{12}, I got the distinct feeling that he was inside those marvelous molecules, moving about each with sure and total understanding of the topography and chemical behaviors of their every part. There are doubtless other material things—physical systems, machines, complex geological sites of fractured, overlayered, and overturned stratigraphy—that physicists, engineers, and geologists similarly know and understand as virtual inhabitants.

Another such case, from plant genetics, is Barbara McClintock's uncanny ability to perceive and understand the complex transpositions of "jumping genes" between the chromosomes of *Zea mays*. Her singular capacity has been termed "a feeling for the organism"—an expression that to some has implied mystical powers or feminine intuition and empathy, but that more plausibly points to her signature practice of observing maize plants, as it were, from inside. In Evelyn Fox Keller's recollection of McClintock's own words, "The 'molecular' revolution in biology was a triumph of the kind of science represented by classical physics. Now, the necessary next step seems to be the reincorporation of the naturalist's approach—an approach that does not press nature with leading questions but *dwells patiently in the variety and complexity of organisms*."[12] The understanding that comes from patient dwelling in something may seem beyond reason, but it is the distinctive empirical reasoning of resident science. How many other Woodwards or McClintocks there may be in the history of science we don't know. Can one have a feeling for atomic nuclei, or cells, or eutrophic ponds, or melting glaciers? Is having "a feeling for" a telltale sign of inside science?

Extending the reach of inside science need not follow lines of subject or discipline. We could equally well proceed by exploring the varieties of human occupation for suggestive analogies. That is what anthropologists did when they lumped themselves together with early modern peddlers and moneylenders in Georg Simmel's category of "stranger." And we have met other examples of that analogical reasoning. Sociologists' concept of hometown sociology envisioned a science built on the common experience of indigeneity—being born and bred in a place.[13] The idea of "commuting" science likewise connects a practice of science with a common mode of suburban or exurban habitation. And would it be too great a stretch to look for inner-city modes of science analogous to the packed, diverse, and intensely interactive life of city blocks that Jane Jacobs celebrated? The high-pressure multidisciplinary ghettos of corporate and state "big" science come to mind—Bell Labs, Los Alamos, NIH Bethesda, CERN, Akademgorodok—or the proliferating island archipelagoes of biotech campuses and silicon valleys.[14] Such studies would have the added benefit of taking the history of situated science across the categorical divide between the physical and life sciences.

The peripatetic ways of life of our restless species are an especially rich source of analogies for inside science. Historians have fruitfully likened scientists' summer sojourns in marine and field stations to middle-class customs of seaside and lakeside vacationing—science as a vacation.[15] An analogy might equally be drawn between survey and collecting expeditions and pastoralists' ancient custom of transhumance—seeking the seasonal necessities of life wherever they were to be found. And analogies might be further sought in the many varieties of agricultural habitation, from the reductive environments of industrial monoculture to the natural disorderly order of swidden or polyculture gardening.[16] Polyculture, in its embrace of context and total knowledge of small locales, is especially suggestive of inside science.

And then there is the more extreme case of the nomadic pastoralists of the inner Asian steppes, who conceived of residence and homeland not as a bounded and mappable geography of place but as a moving set of relationships with the natural and supernatural elements and species on which their lives depended: grasses, marmots, gerbils, horses, and sheep, as well as winds, rain, and spirits.[17] Wherever these were found all together, there was home. This relational conception of inhabiting is a resonant one for thinking about inside science and about place in general. What are places, after all, to the people and the creatures who inhabit them, but webs of social and ecological relationships? And what is inside science but the study of such relationships by yet another species of inhabitant in search of the necessities of life—scientific life in their case—observing on the inside?

This is, I confess, a rather rambling and open-ended way to end a book. Some of my thoughts will seem fanciful and may well be so. Yet isn't that how exploratory forays such as this one should end: not by closing a subject for a generation but by opening it up for those who might come next? Is not a ramble the most appropriate way to give shape and empirical substance to the sprawling and diverse world of inside science?

ACKNOWLEDGMENTS

In the many years it has taken me to write this book, I have greatly benefited from the critical reading, advice, and encouragement of friends and colleagues. Jeremy Vetter and Lynn Nyhart read one or more early drafts, assured me when I needed reassurance, and pointed me in better directions. I am grateful for their friendship and candid counsel. Mary Morgan read several drafts of the sociology chapters and set me right on subjects she knows far better than I ever will. I owe much to our engaging conversations over the years on matters of general and mutual interest. Albert Way generously shared his expert knowledge of wildlife ecology, and Etienne Benson perused selectively and offered useful response. Death deprived me of Riki Kuklick's pungent wisdom on all things ethnographic, but her voice was always in my mind. And I am especially grateful to the two anonymous readers for the University of Chicago Press, who provoked me to read and think more deeply about topics on which I had lazily skimped, and whose judicious and persistent advice made the order and substance of my narrative more cogent and user-friendly. Last but not least, I thank Frances Coulborn Kohler once again for casting her sharp editorial eye over more than one "final" draft. No one does it better.

NOTES

CHAPTER ONE

1. By the "modern" period I mean, roughly, the last two centuries. Distinctions between desituated and situated science, and between science and common life would apply imperfectly if at all to the practices of premodern natural philosophy and natural history, and may prove less apt as well in the churning world of postmodernity that is taking (or losing) shape all around us.

2. I use the term "situated" to mean the natural and human circumstances of observers and observed; however, the term has other and potentially confusing usages. In science studies it typically means the cultural contexts in which scientists operate (ideological, sociopolitical, gender). Donna Haraway, for example, uses "situated knowledges" to mean knowledge "situated" in the gendered bodies of those who produce them. Donna J. Haraway, *Simians, Cyborgs, and Women: The Reinvention of Nature* (New York: Routledge, 1991), 110–11, 188, 198. For a broader overview: Ian C. Jarvie, "Situational logic and its reception," *Philosophy of the Social Sciences* 28 (1998): 365–80; and Adele E. Clarke, *Situational Analysis: Grounded Theory after the Postmodern Turn* (Thousand Oaks, CA: Sage, 2005).

3. General treatments of place and science include David N. Livingstone, *Putting Science in Its Place* (Chicago: University of Chicago Press, 2003), esp. chap. 2; Thomas F. Gieryn, "Three truth spots," *Journal of the History of the Behavioral Sciences* 38 (2002): 113–32; Gieryn, *Truth Spots: How Places Make People Believe* (Chicago: University of Chicago Press, 2018); Richard W. Burkhardt, Jr., "Ethology, natural history, the life sciences, and the problem of place," *Journal of the History of Biology* 32 (1999): 489–508; Robert E. Kohler, "Place and practice in field biology," *History of Science* 40 (2002): 189–210; and Sharon Kingsland, "The role of place in the history of ecology," in *The Ecology of Place: Contributions of Place-Based Research to Ecological Understandings*, ed. Ian Billick and Mary V. Price (Chicago: University of Chicago Press, 2010), 15–39.

4. Bruno Latour and Steven Woolgar, *Laboratory Life: The Construction of Scientific Facts* (Beverly Hills, CA: Sage, 1979; revised edition, Princeton, NJ: Princeton University Press, 1986), esp. chap. 4.

5. Jim Endersby, *A Guinea Pig's History of Biology: The Plants and Animals Who Taught Us the Facts of Life* (London: Heinemann, 2007).

6. Steven Shapin, *The Scientific Life: A Moral History of a Late Modern Vocation* (Chicago: University of Chicago Press, 2008), chaps. 1–2; and Latour and Woolgar, *Laboratory Life* (n. 4), chap. 5.

7. Jonah Lehrer, "The truth wears off: Is there something wrong with the scientific method?," *New Yorker*, 13 Dec. 2010, 52–57.

8. John Gerring, "What is a case study and what is it good for?," *American Political Science Review* 98 (2004): 341–54. As Gerring puts it (353), "There is no virtue, and potentially great harm, in pursuing one approach to the exclusion of the other or in ghettoizing the practitioners of the minority approach."

9. Lehrer, "Truth wears off" (n. 7), quote 57.

10. A group of diverse and suggestive essays on alternatives to universalizing laws is Angela N. H. Creager, Elizabeth Lunbeck, and M. Norton Wise, eds., *Science without Laws: Model Systems, Cases, Exemplary Narratives* (Durham, NC: Duke University Press, 2007).

11. Bruno Latour, "Give me a laboratory and I will move the world," in *Science Observed: Perspectives on the Social Study of Science*, ed. Karen Knorr-Cetina and Michael Mulkay (London: Sage, 1983), 141–70. Latour's account is less empirical history than parable, as is my riff on it.

12. On virtual witnessing: Steven Shapin, "Pump and circumstance: Robert Boyle's literary technology," *Social Studies of Science* 14 (1984): 481–520.

13. Edward O. Wilson, "One giant leap: How insects achieved altruism and colonial life," *BioScience* 58 (2008): 17–25, quote 17.

14. Mary S. Morgan, "The curious case of the prisoner's dilemma: Model situation? Exemplary narrative?," in Creager et al., *Science without Laws* (n. 10), 157–85.

15. Jorge Luis Borges, "On exactitude in science" (1946), in *A Universal History of Infamy* (second edition, New York: Dutton, 1972), 141.

16. Robert E. Kohler, "Labscapes: Naturalizing the lab," *History of Science* 40 (2002): 473–501; Jeremy Vetter, "Rocky Mountain high science: Teaching, research, and nature at field stations," in *Knowing Global Environments: New Historical Perspectives on the Field Sciences*, ed. Vetter (New Brunswick, NJ: Rutgers University Press, 2011), 108–34; Raf de Bont, "Between the laboratory and the deep blue sea: Space issues in the marine stations of Naples and Wimereux," *Social Studies of Science* 39 (2009): 199–227; Stephen Bocking, "Science and spaces in the northern environment," *Environmental History* 12 (2007): 868–94; Deborah R. Coen, "The storm lab: Meteorology in the Austrian Alps," *Science in Context* 22 (2009): 463–856; and Kohler, *Landscapes and Labscapes: Exploring the Lab-Field Border in Biology* (Chicago: University of Chicago Press, 2002), chaps. 4–5 (ecological experiments) and chaps. 7–8 (nature's experiments, "practices of place").

17. Susan M. Pearce, *Museums, Objects, and Collections: A Cultural Study* (Washington, DC: Smithsonian Institution Press, 1992); Robert E. Kohler, "Finders, keepers: Collecting sciences and collecting practice," *History of Science* 45 (2007):

428–54; and Kohler, *All Creatures: Naturalists, Collectors and Biodiversity, 1850–1950* (Princeton, NJ: Princeton University Press, 2006), 149–54 (curators and fieldwork).

18. Warwick Anderson, "Natural histories of infectious disease: Ecological vision in twentieth-century biomedical science," *Osiris* 19 (2004): 39–61; and J. Andrew Mendelsohn, "The microscopist of modern life," *Osiris* 18 (2003): 150–70.

19. Thomas F. Gieryn, "City as truth-spot: Laboratories and field-sites in urban studies," *Social Studies of Science* 36 (2006): 5–38.

20. Gerring, "What is a case study?" (n. 8), quote 346.

21. On the concept of border zones: Kohler, *Landscapes and Labscapes* (n. 16), 14–21.

22. Robert E. Kohler, "Paul Errington, Aldo Leopold, and wildlife ecology: Residential science," *Historical Studies in the Natural Sciences* 41 (2011): 216–54.

23. On case study generally: John Forrester, "If *p*, then what? Thinking in cases," *History of the Human Sciences* 9 (1996): 1–25; Gerring, "What is a case study?" (n. 8); Charles C. Ragin and Howard S. Becker, eds., *What Is a Case? Exploring the Foundations of Social Inquiry* (New York: Cambridge University Press, 1992), esp. Becker, "Cases, causes, conjunctures, stories, and imagery," 205–16. Also Mary S. Morgan, "Case studies: One observation or many? Justification or discovery?," *Philosophy of Science* 79 (2012): 667–77; and Morgan, "Resituating knowledge: Generic strategies and case studies," *Philosophy of Science* 81 (2014): 1012–24.

24. Howard S. Becker, *What about Mozart? What about Murder? Reasoning from Cases* (Chicago: University of Chicago Press, 2014), quote 60. Also Becker, "The art of comparison: Lessons from the master, Everett C. Hughes," in *The Legacy of the Chicago School*, ed. Christopher Hart (Poynton, UK: Midrash, 2010), 185–94.

25. Carlo Ginzburg, "Morelli, Freud, and Sherlock Holmes: Clues and scientific method," *History Workshop* 9 (1980): 5–34. An earlier, shorter version is "Roots of a scientific paradigm," *Theory and Society* 7 (1979): 273–88.

26. Ginzburg, "Morelli, Freud" (n. 25), 7–14, quotes 10 (Freud) and 14 (hunters).

27. Ginzburg, "Morelli, Freud" (n. 25), 20–23, 27–29; and 13, 23 ("method of Zadig").

28. Ginzburg, "Morelli, Freud" (n. 25), 14–16, 20–22 (conjecture), 15–20 (method of Galileo).

29. Gerring, "What is a case study?" (n. 8), quote 349.

30. Ginzburg, "Morelli, Freud" (n. 25), 27–29.

31. Samuel Gorovitz and Alasdair MacIntyre, "Toward a theory of medical fallibility," *Hastings Center Report* 5 (December 1975): 13–23, esp. 15–18. I am grateful to Mary Morgan for directing me to this remarkable essay.

32. Gorovitz and MacIntyre, "Toward a theory" (n. 31), 17.

33. Gorovitz and MacIntyre, "Toward a theory" (n. 31), quote 17.

34. Gorovitz and MacIntyre, "Toward a theory" (n. 31), quote 17.

35. Alfred Schütz, "Common-sense and scientific interpretation of human action," *Philosophy and Phenomenological Research* 14 (1953): 1–37, esp. 28–34, quote 31. I thank Thomas Gieryn for pointing me to Schütz's writings (though we read them quite differently).

36. Schütz, "Common-sense" (n. 35), quote 31.

37. Kohler, "Paul Errington" (n. 22).

38. Everett C. Hughes, "Introduction: The place of field work in social science," in *Field Work: An Introduction to the Social Sciences*, ed. Buford H. Junker (Chicago: University of Chicago Press, 1962), v–xv, at xiv.

39. Georg Simmel, "The stranger" (1908), in Simmel, *On Individuality and Social Forms* (Chicago: University of Chicago Press, 1971), 143–49, esp. 143–46; and Dennison Nash, "The ethnologist as stranger: An essay in the sociology of knowledge," *Southwestern Journal of Anthropology* 19 (1963): 149–67. Nash emphasizes the limitations of what ethnographer-strangers can know about cultures not their own.

40. Charles Elton, *Animal Ecology* (London: Sidgwick and Jackson, 1927), quote 64. Elton was thinking specifically of an animal's place in a food chain or food web, but the thought can be generalized.

41. Lorraine Daston, "On scientific observation," *Isis* 99 (2008): 97–110; and essays in Daston and Elizabeth Lunbeck, eds., *Histories of Scientific Observation* (Chicago: University of Chicago Press, 2011), esp. Daston, "The empire of observation, 1600–1800," 81–113, and Gianna Pomata, "Observation rising: Birth of an epistemic genre, 1500–1650," 45–80. Comparable overviews of observation in nineteenth- and twentieth-century science remain to be written.

42. Daston, "Empire of observation" (n. 41), quote 104.

43. Lorraine Daston and Elizabeth Lunbeck, "Observation observed," in Daston and Lunbeck, *Histories of Scientific Observation* (n. 41), 1–9, at 3–5; Lorraine Daston, email to the author, 28 July 2011.

44. John F. W. Herschel, *A Preliminary Discourse on the Study of Natural Philosophy* (London: Longman, Rees, Orme, Brown, Green, and Longman, 1833), 76–77; and Daston and Lunbeck, "Observation observed" (n. 43), 4–5.

45. Harro Maas, "Sorting things out: The economist as an armchair observer," in Daston and Lunbeck, *Histories of Scientific Observation* (n. 41), 206–29, quote 217; also Maas and Mary S. Morgan, "Observation and observing in economics," *History of Political Economy* 44, supplement (2012): 1–24.

46. T[heodore] C. Schneirla, "The relationship between observation and experimentation in the field study of behavior," *Annals of the New York Academy of Sciences* 51, no. 6 (1950): 1022–44, at 1024–26.

47. Ginzburg, "Morelli, Freud" (n. 25), quote 28.

48. David Carr, "Narrative explanation and its malcontents," *History and Theory* 47 (2008): 19–30, quotes 21–22 (familiar repertoire, reassuring); and Carr, "Narrative and the real world: An argument for continuity," *History and Theory*

25 (1986): 117–31, esp. 117–21. I am grateful to Mary Morgan for directing me to Carr's work.

49. Carr, "Narrative explanation" (n. 48), 22–26. Also Hayden White, "The question of narrative in contemporary historical theory," *History and Theory* 23 (1984): 1–33.

50. Landmarks in the rebirth of narrative history include Lawrence Stone, "The revival of narrative: Reflections on a new old history," *Past & Present* 85 (1979): 3–24; and Peter Burke, "History of events and the revival of narrative," in *New Perspectives on Historical Writing*, ed. Burke (University Park: Pennsylvania State University Press, 1992), 233–48.

51. On the narrative structure of reality: Carr, "Narrative and the real world" (n. 48), 117, 121–22, 125–26, 131; Carr, "Narrative explanation" (n. 48), 20, 28–30; and David Carr, "Getting the story straight: Narrative and historical knowledge," *Poznan Studies in the Philosophy of the Sciences and the Humanities* 41 (1994): 119–33, at 121–23.

52. Carr, "Narrative and the real world" (n. 48), quotes 125 (storytelling, not just embellishment) and 128 (practical, constitutive); and Carr, "Getting the story straight" (n. 51), quote 123 (mode of being). Also Carr, "Narrative explanation" (n. 48), 28–30.

53. Carr, "Narrative explanation" (n. 48), 27–30, quotes 28 (depart from common sense) and 30 (narratives all the way down).

54. Mary S. Morgan and M. Norton Wise, "Narrative science and narrative knowing: Introduction to special issue on narrative science," *Studies in History and Philosophy of Science* 62 (2017): 1–5, at 1 (neglect of narrative).

55. Mary S. Morgan and M. Norton Wise, eds., special issue of *Studies in History and Philosophy of Science* 62 (2017): 1–98; Wise, "On the narrative form of simulations," *Studies in History and Philosophy of Science* 62 (2017): 74–85, at 84 (conferences); and Morgan, "Narrating, ordering, and explanation in the sciences: Historical investigations and perspectives," EU AIG application, June 2015, 4–13, 18–21. I am grateful to Mary Morgan for allowing me to use this document.

56. Mary S. Morgan, "Models, stories and the economic world," *Journal of Economic Methodology* 8 (2001): 361–84; and Morgan, "Curious case" (n. 14); M. Norton Wise, "Science as (historical) narrative," *Erkenntnis* 75 (2011): 349–76; and Wise, "On the narrative form" (n. 55).

57. Wise, "On the narrative form" (n. 55), quote 76.

58. Morgan's list includes, in addition to the proven sciences of natural history, anthropology, and clinical medicine, those that rely on models and simulations; historical sciences (geology, paleontology, evolution); sciences of complexity like ecology, environmental science, and climate science; and sciences that employ case study or counterfactuals. Morgan, "Narrating, ordering, and explanation" (n. 55), 4–13, 18–21.

59. Morgan and Wise, special issue (n. 55), quote 2. On the varied concepts of

relationship and association afforded by narrative: Mary S. Morgan, "Narrative ordering and explanation," *Studies in History and Philosophy of Science* 62 (2017): 86–97, at 86–87.

60. Wise, "Science as (historical) narrative" (n. 56), 372.

CHAPTER TWO

1. Raymond Firth, "Malinowski in the history of social anthropology," in *Malinowski between Two Worlds: The Polish Roots of an Anthropological Tradition*, ed. Roy Ellen, Ernest Gellner, Grażyna Kubica, and Janusz Mucha (Cambridge: Cambridge University Press, 1988), 12–42; and Firth, "Bronislaw Malinowski," in *Totems and Teachers: Perspectives on the History of Anthropology*, ed. Sydel Silverman (New York: Columbia University Press, 1981),101–39.

2. Edmund R. Leach, "The epistemological background to Malinowski's empiricism," in *Man and Culture: An Evaluation of the Work of Bronislaw Malinowski*, ed. Raymond Firth (London: Routledge and Kegan Paul, 1957), 119–37, quote 119. Also Michael W. Young, "Malinowski and the function of culture," in *Creating Culture: Profiles in the Study of Culture* (Sydney: Allen and Unwin, 1987), 124–50, at 124–25, 130–32.

3. Bronislaw Malinowski, *Argonauts of the Western Pacific* (London: George Routledge and Sons, 1922), quote 4. Instances of despondency and escape into novels abound in his private diary. Malinowski, *A Diary in the Strict Sense of the Term* (London: Routledge and Kegan Paul, 1967).

4. Malinowski, *Argonauts* (n. 3), 4–6, quotes 6.

5. George W. Stocking, Jr., "The ethnographer's magic: Fieldwork in British anthropology from Tylor to Malinowski," in *Observers Observed: Essays on Ethnographic Fieldwork*, ed. Stocking (Madison: University of Wisconsin Press, 1983), 70–120; Stocking, *After Tylor: British Social Anthropology, 1888–1951* (Madison: University of Wisconsin Press, 1995); Henrika Kuklick, "After Ishmael: The fieldwork tradition and its future," in *Anthropological Locations: Boundaries and Grounds of a Field Science*, ed. Akhil Gupta and James Ferguson (Berkeley: University of California Press, 1997), 47–65; and Kuklick, *The Savage Within: The Social History of British Anthropology, 1885–1945* (New York: Cambridge University Press, 1991). Also James Clifford, "On ethnographic authority," *Representations* 1, no. 2 (Spring 1983): 118–46.

6. Michael W. Young, "Introduction," in *The Ethnography of Malinowski: The Trobriand Islands, 1915–1918*, ed. Young (London: Routledge, 1979), 1–20, at 1 (right man right place); and Stocking, "Ethnographer's magic" (n. 5), 83–85.

7. Meyer Fortes, "Malinowski and the study of kinship," in Firth, *Man and Culture* (n. 2), 157–88, quote 157.

8. Firth, "Malinowski in the history" (n. 1), 31–32.

9. What follows is drawn from Michael W. Young, *Malinowski: Odyssey of an Anthropologist, 1884–1920* (New Haven, CT: Yale University Press, 2004); Young,

"The intensive study of a restricted area, or why did Malinowski go to the Trobriand Islands?," *Oceania* 55 (1984): 1–26; Young, "The ethnographer as hero: The imponderabilia of Malinowski's everyday life in Mailu," *Canberra Anthropology* 10, no. 2 (1987): 32–50; and Phyllis Kaberry, "Malinowski's contribution to field-work methods and the writing of ethnography," in Firth, *Man and Culture* (n. 2), 71–91.

10. Victoria J. Baker, "Pitching a tent in the native village: Malinowski and participant observation," *Bijdragen tot de Taal-, Land- en Volkenkunde* 143 (1987): 14–24.

11. Stocking, "Ethnographer's magic" (n. 5), 80–85; Stocking, *After Tylor* (n. 5), chap. 3; Kuklick, *Savage Within* (n. 5), chap. 2; and James Urry, "*Notes and Queries on Anthropology* and the development of field methods in British anthropology, 1870–1920," *Proceedings of the Royal Anthropological Institute* (1972): 45–57, at 50.

12. Urry, "*Notes and Queries*" (n. 11); and Stocking, "Ethnographer's magic" (n. 5), 71–73.

13. Sara Sohmer, "The Melanesian mission and Victorian anthropology: A study in symbiosis," in *Darwin's Laboratory: Evolutionary Theory and Natural History in the Pacific*, ed. Roy MacLeod and Philip F. Rehbock (Honolulu: University of Hawaii Press, 1994), 317–38; Niel Gunson, "British missionaries and their contribution to science in the Pacific islands," in MacLeod and Rehbock, *Darwin's Laboratory*, 283–316; Stocking, "Ethnographer's magic" (n. 5), 71–75; and Kuklick, "After Ishmael" (n. 5), 54–59.

14. Malinowski, *Argonauts* (n. 3), 17–18; and Young, *Malinowski: Odyssey* (n. 9), 380–83.

15. W. H. R. Rivers, "Report on anthropological research outside America," in *Reports upon the Present Conditions and Future Needs of the Science of Anthropology*, ed. Rivers, A. E. Jenks, and S. G. Morley, Publication 200 (Washington, DC: Carnegie Institution of Washington, 1913): 5–28, at 8–10.

16. Bronislaw Malinowski, *Coral Gardens and Their Magic* (London: George Allen and Unwin, 1935; reprint, Bloomington: University of Indiana Press, 1965), 326; and Malinowski, *Argonauts* (n. 3), 5–6.

17. Kuklick, "After Ishmael" (n. 5); Henrika Kuklick, "The color blue: From research in the Torres Strait to an ecology of human behavior," in MacLeod and Rehbock, *Darwin's Laboratory* (n. 13), 339–67, at 341–42; Stocking, "Ethnographer's magic" (n. 5); and Stocking, *After Tylor* (n. 5), chap. 3. Also Anita Herle and Sandra Rouse, eds., *Cambridge and the Torres Strait* (Cambridge: Cambridge University Press, 1998).

18. Robert E. Kohler, "Finders, keepers: Collecting sciences and collecting practice," *History of Science* 45 (2007): 428–54; and Kohler, *All Creatures: Naturalists, Collectors, and Biodiversity, 1850–1950* (Princeton, NJ: Princeton University Press, 2006), 107–23 (curators and collecting).

19. On historicism: Stocking, *After Tylor* (n. 5); and Kuklick, *Savage Within* (n. 5).

20. Urry, "*Notes and Queries*" (n. 11), 48. Also Stocking, *After Tylor* (n. 5), 115–23.

21. A. C. Haddon, "Presidential address to Section H," in *Report of the Seventy-Fifth Meeting of the British Association for the Advancement of Science* (London: John Murray, 1905), 511–27, at 525; and Haddon, "A plea for the investigation of biological and anthropological distributions in Melanesia," *Geographical Journal* 28 (1906): 155–63, at 156–57, quote 157.

22. Haddon, "Plea" (n. 21), 156–57; George W. Stocking, Jr., "The intensive study of limited areas—toward an ethnographic context for the Malinowski innovation," *History of Anthropology Newsletter* 6, no. 2 (1979): 9–12. Stocking writes (9) that Malinowski's work "was as much the culmination of a Torres straits ethnographic tradition as it was the starting point of a modern functionalist one."

23. Stocking, *After Tylor* (n. 5), 116–20. Landtman lived mainly at a missionary station, relied on formal interview of paid informants, and never learned the native language.

24. Rivers, "Anthropological research" (n. 15), 6–8, 13–14, quote 6.

25. Rivers, "Anthropological research" (n. 15), 18–20.

26. W. H. R. Rivers, "The genealogical method of anthropological inquiry," *Sociological Review* 3 (1910): 1–12, at 9–11, quote 9 (formulate laws); and Rivers, "A general account of method," in *Notes and Queries on Anthropology*, 4th edition, ed. Barbara Freire-Marreco and John Linton Myers (London: Royal Anthropological Institute, 1912), 108–28, at 119–22, quote 119 (the very instrument). Also Stocking, "Ethnographer's magic" (n. 5), 187–88.

27. Malinowski, *Argonauts* (n. 3), 12–13, quote 12. On field techniques: Rivers, "General account" (n. 26); Stocking, "Ethnographer's magic" (n. 5), 89–93; and Stocking, *After Tylor* (n. 5), chap. 3.

28. Robert R. Marett, unpublished foreword, c. 1912, quoted in Wilson D. Wallis, "Anthropology in England early in the present century," *American Anthropologist* 59 (1957): 781–90, at 789–790.

29. Marett, unpublished foreword (n. 28), quote 790. On Marett: Young, *Malinowski: Odyssey* (n. 9), 170. Although Marett believed in separating empirical work in the field from intellectual work at home, he was an avid promoter of fieldwork and by 1914 had sent two of his own students into the field.

30. Kuklick, "Color blue" (n. 17), 347.

31. Young, *Malinowski: Odyssey* (n. 9), 155–56.

32. Robert Thornton, " 'Imagine yourself set down . . .': Mach, Frazer, Conrad, Malinowski and the role of imagination in ethnography," *Anthropology Today* 1, no. 5 (1985): 7–14, quotes 13–14. Also Young, *Malinowski: Odyssey* (n. 9), 83–88.

33. Malinowski, *Argonauts* (n. 3), quote 100.

34. Young, *Malinowski: Odyssey* (n. 9), 67–72.

35. Principal sources on Malinowski's early life include Young, *Malinowski: Odyssey* (n. 9), chaps. 1–3; Robert Thornton and Peter Skalnik, eds., *The Early*

Writings of Bronislaw Malinowski (Cambridge: Cambridge University Press, 1993); Thornton and Skalnik, "Introduction: Malinowski's reading, writing, 1902–1914," in *Early Writings*, 1–64; Thornton, "'Imagine yourself" (n. 32); and Skalnik, "Bronislaw Kaspar Malinowski and Stanislaw Ignacy Witkiewicz: Science versus art in the conceptualization of culture," in *Fieldwork and Footnotes: Studies in the History of European Anthropology*, ed. Han F. Vermeulen and Arturo Alvarez Roldán (London: Routledge, 1995), 129–42. On the Polish roots of Malinowski's thinking: Ellen et al., *Malinowski between Two Worlds* (n. 1); appendix 3 is a comprehensive bibliography of Malinowski's published works to 1986.

36. Young, *Malinowski: Odyssey* (n. 9), chap. 5 (student years). On Malinowski and Mach: Young, *Malinowski: Odyssey* (n. 9), 82–89; Thornton and Skalnik, "Introduction" (n. 35), 5–8, 16–26, 26–38; and Bronislaw Malinowski, "On the principle of the economy of thought," in Thornton and Skalnik, *Early Writings* (n. 35), 89–115 (dissertation).

37. Young, *Malinowski: Odyssey* (n. 9), 82; Thornton and Skalnik, "Introduction" (n. 35), 5.

38. Thornton and Skalnik, "Introduction" (n. 35), 32–38; Young, *Malinowski: Odyssey* (n. 9), 85–88.

39. Young, *Malinowski: Odyssey* (n. 9), 138 (heat physics), 130 (path of least resistance).

40. Young, *Malinowski: Odyssey* (n. 9), 140–41, 151, 225–26.

41. Young, *Malinowski: Odyssey* (n. 9), 137–40.

42. Young, *Malinowski: Odyssey* (n. 9), 137, 144–45, 146–47.

43. Young, *Malinowski: Odyssey* (n. 9), quote 137.

44. Young, *Malinowski: Odyssey* (n. 9), 140–41, 147, 176–82; and Bronislaw Malinowski, *The Family among the Australian Aborigines: A Sociological Study* (London: Hodder and Stoughton, 1913). Thornton and Skalnik make a similar point about another of Malinowski's early essays, from 1912, which they view as his first sociological analysis of ethnological material. Thornton and Skalnik, "Introduction" (n. 35), 49–51. Also Malinowski, "Tribal male associations of the Australian aborigines," *Bulletin International de L'Academie des Science de Cracovie*, Classe d'Histoire et de Philosophie, nos. 4–6 (1912): 56–63. Other apprentice essays in the same mode include Malinowski, "Totemism and exogamy," and Malinowski, "The economic aspects of the *Intichiuma* ceremonies" (1912), both in *Early Writings*, ed. Thornton and Skalnik (n. 35), 123–99, 209–27.

45. Quoted in Young, *Malinowski: Odyssey* (n. 9), 154. The unpublished diary that Young sources here dates from September 1911.

46. Bronislaw Malinowski, *Crime and Custom in Savage Society* (London: Kegan Paul, Trench, Trubner, 1926), 40–41, quote 40; Malinowski, *The Sexual Life of Savages in North-Western Melanesia* (New York: Horace Liveright, 1929), xxvi; and Malinowski, *Coral Gardens* (n. 16), 452–82, appendix 2.

47. The following account of Malinowski's three field trips is drawn from Young,

"Intensive study" (n. 9), 2–11; Young, *Malinowski: Odyssey* (n. 9), chaps. 16–18; and Stocking, "Ethnographer's magic" (n. 5), 97–104.

48. Young, "Ethnographer as hero" (n. 9), 39–40 (time in field); and Young, *Malinowski: Odyssey* (n. 9), 388–89, 393–94, 494–502 (respites with Europeans).

49. Malinowski, *Diary in the Strict Sense* (n. 3), 97, entry 28 Feb. 1915. It is unclear whether "old fears" refers to Malinowski's crisis of confidence in Leipzig, in addition to the self-doubts of his first weeks in New Guinea.

50. Young, *Malinowski: Odyssey* (n. 9), 245–46.

51. Young, "Intensive study" (n. 9), 18–19.

52. Young, *Malinowski: Odyssey* (n. 9), 312–13, 325–27, 367, 377, 437–42.

53. Bronislaw Malinowski, "The natives of Mailu," *Transactions and Proceedings of the Royal Society of South Australia* 39 (1915): 494–706, reprinted as *Malinowski among the Magi: The Natives of Mailu*, ed. Michael W. Young (London: Routledge, 1988).

54. Young, *Malinowski: Odyssey* (n. 9), 322–33; and Stocking, "Ethnographer's magic" (n. 5), 89–93.

55. Malinowski, *Coral Gardens* (n. 16), quote 326.

56. Young, *Malinowski: Odyssey* (n. 9), 468–71.

57. Young, *Malinowski: Odyssey* (n. 9), 383–97, esp. 390–97 (arrival and reception, and initial surveys).

58. On "imponderabilia": Malinowski, *Argonauts* (n. 3), 18–19; Kaberry, "Malinowski's contribution" (n. 9), 85–86; Young, "Ethnographer as hero" (n. 9), 39–40; and Young, *Malinowski: Odyssey* (n. 9), 397–98.

59. For example, Malinowski, *Coral Gardens* (n. 16), 317–20, quote 317 ("charters").

60. Accounts of Malinowski's practice of participant observation include Malinowski, *Argonauts* (n. 3), 1–25; Stocking, "Ethnographer's magic" (n. 5), 93–106; Young, *Malinowski: Odyssey* (n. 9), 394–405, 502–13; Baker, "Pitching a tent" (n. 10); and Kaberry, "Malinowski's contribution" (n. 9). Also Audrey I. Richards, "The development of field work methods in social anthropology," in *The Study of Society: Methods and Problems*, ed. Frederic C. Bartlett, M. Ginsberg, E. J. Lindgren, and R. H. Thouless (London: Routledge and Kegan Paul, 1939), 272–316.

61. Young, *Malinowski: Odyssey* (n. 9), 390–94 (arrival, tent); Malinowski, *Coral Gardens* (n. 16), 453; Malinowski, *Argonauts* (n. 3), 23–24 (language); and Bronislaw Malinowski, "Baloma: Spirits of the dead in the Trobriand Islands," *Journal of the Royal Anthropological Institute* 46 (1916): 354–430, at 427 (Bagido'u).

62. Malinowski, *Argonauts* (n. 3), 13–17; and Malinowski, *Diary in the Strict Sense* (n. 3), 290.

63. Malinowski, "Baloma" (n. 61), 418–30; and Malinowski, *Argonauts* (n. 3), 17–22.

64. Jürgen Haffer, "The history of species concepts and species limits in ornithology," *Bulletin of the British Ornithological Club*, supplement 112A (1992): 107–58; and Kohler, *All Creatures* (n. 18), chap. 6.

65. Young, *Malinowski: Odyssey* (n. 9), 397–402, quotes 397 (surreal), 398 (kaleidoscopic), and 401 (trivial).

66. Malinowski to "C.H.W. [Camilla Wedgwood]," 8 May 1933, quoted in Firth, "Malinowski in the history" (n. 1), 29.

67. Bronislaw Malinowski, "Fishing and fishing magic in the Trobriand Islands," *Man* 18 (1918): 87–92.

68. Malinowski, *Crime and Custom* (n. 46), 120–21, quote 120.

69. Malinowski, *Crime and Custom* (n. 46), quote 121.

70. Malinowski, *Crime and Custom* (n. 46), 127.

71. Malinowski, *Argonauts* (n. 3), 17–18.

72. Malinowski, *Argonauts* (n. 3), 7–8, quote 7.

73. Malinowski, "Baloma" (n. 61), quote 428; and Malinowski, *Diary in the Strict Sense* (n. 3), 288 (walking with children).

74. Malinowski, *Argonauts* (n. 3), quote 8; Malinowski, *Coral Gardens* (n. 16), 326. Also Stocking, "Ethnographer's magic" (n. 5), 101–2; and Young, *Malinowski: Odyssey* (n. 9), 404.

75. Malinowski, *Argonauts* (n. 3), quote 21; and Malinowski, *Coral Gardens* (n. 16), 326.

76. Firth, "Bronislaw Malinowski" (n. 1), quote 124.

77. Malinowski, "Baloma" (n. 61), 380.

78. Malinowski, *Argonauts* (n. 3), 479, 16; and Malinowski, *Diary in the Strict Sense* (n. 3), 234, 245 (left behind).

79. Malinowski's diary is full of self-reproach for lecherous feelings toward native (and European) women; and when he once pawed a native woman, he guiltily regretted it afterward. Malinowski, *Diary in the Strict Sense* (n. 3), 256 (pawing).

80. Malinowski, "Baloma" (n. 61), 385 (garden magic).

81. Ruth Fink, "Techniques of observation and their social and cultural limitations," *Mankind* 5, no. 2 (1955): 60–68, 62; and Baker, "Pitching a tent" (n. 10), 18–19.

82. Young, *Malinowski: Odyssey* (n. 9), 528; and Young, "Introduction" (n. 6), 16. Another nickname was Tosemwana, meaning a man who takes on an identity not his own: a performer, mimic, or foolish poseur. Young, *Malinowski: Odyssey* (n. 9), 510.

83. Young, *Malinowski: Odyssey* (n. 9), 397.

84. Young, *Malinowski: Odyssey* (n. 9), 389 (dislike); on Bellamy's efforts to improve the island and its inhabitants, 382–93.

85. Young, *Malinowski: Odyssey* (n. 9), 357–58; on Saville and his work as an ethnographer, 329–34.

86. Malinowski, *Diary in the Strict Sense* (n. 3), quote 167 (dogs). On the critical responses to the diary: George W. Stocking, Jr., "Empathy and antipathy in the heart of darkness," *Journal of the Behavioral Sciences* 4 (1968): 189–94, esp. 190–91; Young, *Malinowski: Odyssey* (n. 9), 404–5; Baker, "Pitching a tent" (n. 10), 19; and Hortense Powdermaker, "Further reflections on Lesu and Malinowski's diary,"

Oceania 40 (1970): 344–47. Malinowski's occasional "hatred" of natives was, I think, more vexation when they failed to perform as promised when he depended on them, especially as informants and in transport (e.g., *Diary in the Strict Sense* [n. 3], 261–62). He expressed the same "hatred" of missionaries and resident whites when they failed him.

87. Malinowski, *Coral Gardens* (n. 16), quote 376–77.

88. Malinowski, *Diary in the Strict Sense* (n. 3), quote 119.

89. Malinowski to Sir James Frazer, 4 Oct. 1917, quoted in Young, *Malinowski: Odyssey* (n. 9), 475–76.

90. Young, "Intensive Study" (n. 9), 17–19, quote 19.

91. The following is based on Malinowski, "Baloma" (n. 61), 402–18; Malinowski, *Sexual Life* (n. 46), 164–202; Malinowski, "Natives of Mailu" (n. 53), 177–78; and Young, *Malinowski: Odyssey* (n. 9), 431–32.

92. Malinowski, "Baloma" (n. 61), quote 411.

93. Malinowski, "Baloma" (n. 61), 411–15; Malinowski, *Sexual Life* (n. 46), 179–89.

94. Malinowski, "Baloma" (n. 61), 370–79.

95. Malinowski, "Baloma" (n. 61), 378–84.

96. Malinowski, "Baloma" (n. 61), quote 381–82.

97. Malinowski, "Baloma" (n. 61), quotes 382 and 370.

98. Malinowski, "Natives of Mailu" (n. 53), 213–19, 254–57; Malinowski, *Coral Gardens* (n. 16), 323.

99. Malinowski, *Coral Gardens* (n. 16), quote 324.

100. Malinowski, *Coral Gardens* (n. 16), 324–30.

101. Malinowski, *Coral Gardens* (n. 16), 321–22 (reader experiment).

102. Malinowski, *Coral Gardens* (n. 16), 454–57, quote 456.

103. Malinowski, *Coral Gardens* (n. 16), 331–40, 453–57, chap. 12. On Malinowski's similar analysis of the kula trade: Malinowski, *Argonauts* (n. 3), esp. 509–18.

104. Malinowski, *Crime and Custom* (n. 46), 14–16 (old views), 19–23, 28–31, 40–42, 58 (new view).

105. Malinowski, *Crime and Custom* (n. 46), 100–110, esp. 102–5 (confrontation).

106. Malinowski, *Crime and Custom* (n. 46), quotes 105 and 107.

107. Malinowski, *Crime and Custom* (n. 46), quote 121. For another dramatic and fatal encounter of law and custom, this one concerning a young couple and the laws of exogeny and incest, see *Crime and Custom*, 35–38, 76–84.

108. Malinowski, *Crime and Custom* (n. 46), 76–84; also 35–38 (matrilineal system).

109. Leach, "Epistemological background" (n. 2), 120.

110. Talcott Parsons, "Malinowski and the theory of social systems," in Firth, *Man and Culture* (n. 2), 53–70, quote 53.

111. Clifford Geertz, "Thick description: Toward an interpretive theory of culture," in *The Interpretation of Cultures: Selected Essays by Clifford Geertz* (New York: Basic, 1973), 3–30.

112. Johann Wolfgang von Goethe, "Ueber Naturwissenschaft im allgemeinen Einzelne Betrachtungen und Aphorismen," in *Schriften zur allgemeinen Naturlehre, Geologie und Mineralogie*, ed. Wolf von Engelhardt und Manfred Wenzel, in *Sämtliche Werke, Briefe, Tagebücher und Gespräche*, part 1, vol. 25 (Frankfurt am Main: Deutscher Klassiker Verlag, 1989), 113 (aphorism 110).

113. Adam Kuper, *Anthropologists and Anthropology: The British School, 1922–1972* (London: Allen Lane, 1973), quote 31. Also Firth, "Bronislaw Malinowski" (n. 1), 103–4.

114. Malinowski, *Argonauts* (n. 3), 9 (molding); Malinowski, *Coral Gardens* (n. 16), 317–21 ("invisible facts"); and Malinowski, *Sexual Life* (n. 46), 384–85.

115. Malinowski, "Baloma" (n. 61), 418–21; and Malinowski, *Crime and Custom* (n. 46), 125–26.

116. Thornton and Skalnik, "Introduction" (n. 35), 37.

117. Bronislaw Malinowski, "Sir James George Frazer: A biographical appreciation," in *A Scientific Theory of Culture and Other Essays* (Chapel Hill: University of North Carolina Press, 1944), 179–221, quotes 184–85. Also Stocking, "Ethnographer's magic" (n. 5), 93–97; Young, *Malinowski: Odyssey* (n. 9), 227–32, 241–45, 475; and Thornton, "Imagine yourself" (n. 32), 8, 10–11.

118. Malinowski, "Sir James George Frazer" (n. 117), quote 212.

119. Thornton and Skalnik, "Introduction" (n. 35), 37–38.

CHAPTER THREE

1. Eduard C. Lindeman, *Social Discovery: An Approach to the Study of Functional Groups* (New York: Republic, 1924), 177–200, quote 192. Lindeman's phrasing suggests that the term was not his invention but a current usage.

2. Jennifer Platt, "The development of the 'participant observation' method in sociology: Origin myth and history," *Journal of the History of the Behavioral Sciences* 19 (1983): 379–93, esp. 384–89, quote 389. Platt, *A History of Sociological Research Methods in America, 1920–1960* (Cambridge: Cambridge University Press, 1996) remains the best study of the subject.

3. On empirical improvement in sociology: Anthony Oberschall, "The institutionalization of American sociology," in *The Establishment of Empirical Sociology*, ed. Oberschall (New York: Harper and Row, 1967), 187–251; and Craig Calhoun, ed., *Sociology in America: A History* (Chicago: University of Chicago Press, 2007). Useful primary sources include Warner E. Gettys, "The field and problems of community study," in *The Fields and Methods of Sociology*, ed. Luther L. Bernard (New York: Ray Long and Richard Smith, 1934), 67–82; and Jesse Bernard, "The history and prospects of sociology in the United States," in *Trends in American Sociology*, ed. George A. Lundberg, Read Bain, and Nels Anderson (New York: Harper, 1929), 1–71.

4. On Columbia sociology: Stephen P. Turner, "The world of the academic quantifiers: The Columbia University family and its connections," in *The Social Survey in Historical Perspective*, ed. Martin Bulmer, Kevin Bales, and Kathryn Kish Sklar (Cambridge: Cambridge University Press, 1991), 269–90; and Robert C. Bannister, *Sociology and Scientism: The American Quest for Objectivity, 1880–1940* (Charlottesville: University of North Carolina Press, 1987), 64–86 (Giddings). On Chicago: Bulmer, *The Chicago School of Sociology: Institutionalization, Diversity, and the Rise of Sociological Research* (Chicago: University of Chicago Press, 1984); and Robert E. L. Faris, *Chicago Sociology, 1920–1932* (Chicago: University of Chicago Press, 1970).

5. James M. Williams, *An American Town: A Sociological Study* (New York: James Kempster, 1906), quotes 9–10.

6. Albion W. Small and George E. Vincent, *An Introduction to the Study of Society* (New York: American Book, 1894), 15–18, 99–168, 367–68.

7. Useful points of entry into the large literature on social survey include Michael Gordon, "The social survey movement and sociology in the United States," *Social Problems* 21 (1973): 284–98; and essays in Bulmer et al., *Social Survey in Historical Perspective* (n. 4), esp. Martin Bulmer, Kevin Bales, and Kathryn Kish Sklar, "The social survey in historical perspective," 1–48, and Bulmer, "The decline of the social survey movement and the rise of American empirical sociology," 291–315. Also Allen H. Eaton and Shelby M. Harrison, *A Bibliography of Social Surveys* (New York: Russell Sage Foundation, 1930), esp. Harrison, "Development and spread of social surveys," xi–xlviii.

8. Harrison, "Development and spread" (n. 7), xxiv–xxvii.

9. On survey practice: Bulmer, "Decline of the social survey" (n. 7), esp. 292–97; and Gordon, "Social survey movement" (n. 7), esp. 286–93.

10. Martin Bulmer, "W. E. B. Du Bois as a social investigator: *The Philadelphia Negro* 1899," in Bulmer et al., *Social Survey* (n. 4), 170–88; W. E. B. Du Bois, *The Philadelphia Negro: A Social Study* (Philadelphia: Ginn, 1899); reprinted with introduction by Elijah Anderson (Philadelphia: University of Pennsylvania Press, 1996). Du Bois lived for the duration in the Seventh Ward and personally carried out the many hundreds of interviews of residents, but he spent on average only twenty to twenty-five minutes in each household and reduced observational data to standard categories for statistical treatment.

11. Harrison, "Development and spread" (n. 7), xxv–xxvi.

12. Useful histories of social work in the United States include Roy Lubove, *The Professional Altruist* (Cambridge, MA: Harvard University Press, 1965); and Stanley Wenocour and Michael Reisch, *From Charity to Enterprise: The Development of American Social Work in a Market Economy* (Urbana: University of Illinois Press, 1984). On the fading away of the social survey movement: Bulmer, "Decline of the social survey" (n. 7), 309–11; and Jean M. Converse, *Survey Research in the United States: Roots and Emergence, 1890–1960* (Berkeley: University of California Press, 1987).

13. Allen F. Davis, *Spearheads for Reform: The Social Settlements and the Progressive Movement, 1890–1914* (New York: Oxford University Press, 1967); Jane Addams, *Twenty Years at Hull-House, with Autobiographical Notes* (New York: Macmillan, 1910); and Rivka Shpak Lissak, *Pluralism and Progressives: Hull-House and the New Immigrants, 1890–1919* (Chicago: University of Chicago Press, 1989).

14. For example, Annie M. MacLean, *Wage Earning Women* (New York: Macmillan, 1910), 100–115 (days of picking hops); MacLean, "Two weeks in department stores," *American Journal of Sociology* 4 (1899): 721–41; and V[irginia?] Fish, "Annie Marion MacLean: A neglected part of the Chicago school," *Journal of the History of Sociology* 3 (1981): 43–62.

15. Small and Vincent, *Introduction* (n. 6), 15–18, 99–168, 367–68.

16. Faris, *Chicago Sociology* (n. 4), 135–50 (list of MA and PhD theses).

17. On community study: Jesse F. Steiner, "An appraisal of the community movement," *Publications of the American Sociological Association* 23 (1929): 15–29, at 16–17; and Steiner, *The American Community in Action* (New York: Henry Holt, 1928), 3–8. Also Arthur E. Wood, "The place of the community in sociological studies," *Publication of the American Sociological Society* 22 (1928): 14–25, at 21–22. Wood endorsed student hometown sociology but did not say if he used it in his own teaching.

18. John Lewis Gillin, "The application of the social survey to small communities," *Publication of the American Sociological Society* 6 (1911): 70–81, quote 70.

19. Ernest W. Burgess, "The social survey: A field for constructive service by departments of sociology," *American Journal of Sociology* 21 (1916): 492–500.

20. Bulmer, "Decline of the social survey" (n. 7), 305–11; Maurice J. Karpf, "The relation between sociology and social work," *Journal of Social Forces* 3 (1925): 1–8; James H. S. Bossard, "The functions and limits of social work as viewed by a sociologist," in Luther L. Bernard, *Fields and Methods of Sociology* (n. 3), 218–31; and Patricia Lengermann and Gillian Niebrugge, "Thrice told: Narratives of sociology's relation to social work," in Calhoun, *Sociology in America* (n. 3), 63–114.

21. Bulmer, "Decline of the social survey" (n. 7), quote 307.

22. Bulmer, "Decline of the social survey" (n. 7), 307.

23. Thomas Jesse Jones, *The Sociology of a New York City Block* (New York: Columbia University Press, 1904); Williams, *American Town* (n. 5); John L. Gillin, *The Dunkers: A Sociological Interpretation* (New York: Columbia University Press, 1906); Warren H. Wilson, *Quaker Hill: A Sociological Study* (New York: Columbia University Press, 1907); Howard B. Woolston, *A Study of the Population of Manhattanville* (New York: Columbia University Press, 1909); and Newell L. Sims, *A Hoosier Village: A Sociological Study with Special Reference to Social Causation* (New York: Columbia University Press, 1912).

24. Williams, *American Town* (n. 5), 9–10; Sims, *Hoosier Village* (n. 23), 11–17; and Wilson, *Quaker Hill* (n. 23), 1–4.

25. Woolston, *Manhattanville* (n. 23), 5–10. Returning two years later, Woolston

found that the community had lost its distinctive village character and become just another urban neighborhood.

26. Jones, *New York City Block* (n. 23), 10–14, quotes 10, 135 (Jones's varied occupations). Head workers supervised settlements' resident workers.

27. Jones, *New York City Block* (n. 23), 7.

28. Bulmer, *Chicago School* (n. 4), 23–24, 33, 37–39; and Steven J. Diner, "Department and discipline: The Department of Sociology at the University of Chicago, 1892–1920," *Minerva* 13 (1975): 514–53.

29. Charles J. Bushnell, *The Social Problem at the Chicago Stock Yards* (Chicago: University of Chicago Press, 1902); and John M. Gillette, *Culture Agencies of a Typical Manufacturing Group: South Chicago* (Chicago: University of Chicago Press, 1901).

30. Bushnell, *Social Problem* (n. 29), xiv–xv.

31. Luther L. Bernard, "Some historical and recent trends of sociology in the United States," *Southwestern Political and Social Science Quarterly* 9 (1928–29): 264–93, quote 291.

32. Bulmer, *Chicago School* (n. 4), 208–24.

33. On the Chicago school: Bulmer, *Chicago School* (n. 4), esp. 34–39 (founding); Faris, *Chicago Sociology* (n. 4); and Platt, *History of Sociological Research* (n. 2). Ruth Shonle Cavan, "The Chicago school of sociology, 1918–1933," *Urban Life* 11 (1983): 407–20, is a participant's insightful recollection.

34. Bulmer, *Chicago School* (n. 4), 94.

35. Herbert Blumer, "Ernest W. Burgess, 1886–1966," *American Sociologist* 2 (1967): 103–4, quote 104.

36. On Thomas: William I. Thomas, "Life history," *American Journal of Sociology* 79 (1973): 247–50; Mary Jo Deegan and Jon S. Burger, "W. I. Thomas and social reform: His work and writings," *Journal of the History of the Behavioral Sciences* 17 (1981): 114–25; and Bulmer, *Chicago School* (n. 4), chap. 4, and passim.

37. Robert E. Park, "Notes on the origin of the society for social research," *Journal of the History of the Behavior Sciences* 18 (1982): 332–57, quoted in Bulmer, *Chicago School* (n. 4), 63.

38. On Park: Fred H. Matthews, *Quest for an American Sociology: Robert E. Park and the Chicago School* (Montreal: McGill-Queens University Press, 1977); Winifred Raushenbush, *Robert E. Park: Biography of a Sociologist* (Durham, NC: Duke University Press, 1979); Everett C. Hughes, ed., *The Collected Papers of Robert Ezra Park*, 3 volumes (Glencoe, IL: Free Press, 1950–55); Robert E. Park, "An autobiographical note," in Park, *Race and Culture* (Glencoe, IL: Free Press, 1950), v–ix; and Park, "Notes on the origin" (n. 37), 336–39.

39. Robert E. Park, "News as a form of knowledge," *American Journal of Sociology* 45 (1940): 669–86, quote 670.

40. Park, "Notes on the origin" (n. 37), quote 338; and Bulmer, *Chicago School* (n. 4), 223.

41. Robert H. Wiebe, *The Search for Order, 1877–1920* (New York: Hill and Wang, 1967) remains a suggestive overview of the transformation of a society of small towns to one of industrial cities.

42. Bulmer, "Decline of the social survey" (n. 7), 300–303; Martin Bulmer, "Charles S. Johnson, Robert E. Park, and the research methods of the Chicago Commission on Race Relations, 1920–1922: An early experiment in applied social research," *Ethnic and Racial Studies* 4 (1981): 289–306; and Matthews, *Quest for an American Sociology* (n. 38), 112–14 (Pacific survey), 176–78 (Chicago survey).

43. Matthews, *Quest for an American Sociology* (n. 38), 8–11, quote 11; and Bulmer, *Chicago School* (n. 4), 68.

44. Bulmer, *Chicago School* (n. 4), 89–91; Faris, *Chicago Sociology* (n. 4), 37–40, 83; and Ernest W. Burgess, "Statistics and case studies as methods of social research," *Sociology and Social Research* 12 (1927): 103–20, at 115.

45. Robert E. Park, "The city: Suggestions for the investigation of human behavior in the city environment," *American Journal of Sociology* 29 (1915): 577–612; Park and Ernest W. Burgess, *Introduction to the Science of Sociology* (Chicago: University of Chicago Press, 1921); and Park, "Notes on the origin" (n. 37), 338. The revised version of Park's essay, in Park, Burgess, and Roderick D. McKenzie, *The City* (Chicago: University of Chicago Press, 1925, reprinted 1967), 1–46, is framed by newly added ecological and geographical concepts and is no guide to Park's thinking in 1915. Pierre Lannoy, "When Robert E. Park was (re)writing 'The City': Biography, the social survey, and the science of sociology," *American Sociologist* 35, no. 1 (2004): 34–62.

46. For example, Robert E. Park, "The city as a social laboratory," in *Chicago: An Experiment in Social Science Research*, ed. Thomas V. Smith and Leonard D. White (Chicago: University of Chicago Press, 1929), 1–19; and Bulmer, *Chicago School* (n. 4), quotes 92–93 (Burgess letter).

47. Thomas F. Gieryn, "City as truth-spot: Laboratories and field-sites in urban studies," *Social Studies of Science* 36 (2006): 5–38.

48. My main source is the bibliography of MA and PhD theses to 1935 in Faris, *Chicago Sociology* (n. 4), 135–50; and Lee Harvey, *Myths of the Chicago School of Sociology* (Aldershot, UK: Avebury, 1987), 284–93, appendix 5 (PhD dissertations to 1950). Because Harvey's list omits MA theses, the two data sets do not, strictly speaking, combine. Production of community studies and case histories dropped off in the mid-1930s as funding for student fieldwork dried up and interest turned from ethnicity and assimilation to the social effects of the Great Depression. Cavan, "Chicago school" (n. 33), 417.

49. The following overview is drawn mainly from Bulmer, *Chicago School* (n. 4), 90–91, 95–98 (practices), 109–28 (Park and Burgess), 136–50 (funding).

50. The words are Park's as recalled by Howard Becker, cited in Bulmer, *Chicago School* (n. 4), 97.

51. Bulmer, *Chicago School* (n. 4), 136–50; and Martin Bulmer, "The early in-

stitutional establishment of social science research: The Local Community Research Committee at the University of Chicago," *Minerva* 18 (1980): 51–110.

52. Bulmer, *Chicago School* (n. 4), 90–91, quote 91; and Park, "Notes on the origin" (n. 37), 338.

53. Albion Small's 1922 plan for a social science institute called for senior faculty doing research, with students helping with the "hack work" of organizing materials. Students would learn research by watching their seniors and by "occasionally taking part in it." Bulmer, *Chicago School* (n. 4), 131–34.

54. Bulmer, *Chicago School* (n. 4), 118–22; Vivien M. Palmer and Ernest W. Burgess, *Social Backgrounds of Chicago's Local Communities* (Chicago: Local Community Research Committee of the University of Chicago), 1930; and Palmer, *Field Studies in Sociology: A Student's Manual* (Chicago: University of Chicago Press, 1928). Also Palmer, "Field studies for introductory sociology," *Journal of Applied Sociology* 10 (1926): 341–48 (organization of undergraduates' fieldwork). Scholars have searched for biographical information on Palmer but found nothing. Christopher Hart, "Introduction," in *The Legacy of the Chicago School*, ed. Hart (Poynton, UK: Midrash, 2010), 1–43, at 1n2.

55. Howard S. Becker, "Introduction," in Clifford R. Shaw, *The Jack-Roller: A Delinquent Boy's Own Story* (Chicago: University of Chicago Press, 1930; new edition, 1966), v–xviii, vii–viii.

56. Bulmer, *Chicago School* (n. 4), 102–6, quote 104; also Shane Blackman, "The 'ethnographic mosaic' of the Chicago school: Critically locating Vivien Palmer, Clifford Shaw and Frederic Thrasher's research methods in contemporary reflexive sociological interpretation," in Hart, *Legacy of the Chicago School* (n. 54), 195–216.

57. Platt, "'Participant observation' method" (n. 2); and Jennifer Platt, "The Chicago school and firsthand data," *History of the Human Sciences* 7 (1994): 57–80.

58. Bulmer, *Chicago School* (n. 4), 89–108, 100–101 (hallmark); Platt, "'Participant observation' method" (n. 2); Jennifer Platt, "'Case study' in American methodological thought," *Current Sociology* 40 (1992): 17–48; and Platt, "Chicago school and firsthand data" (n. 57).

59. Cavan, "Chicago school" (n. 33), quote 415; and Mary Jo Deegan, "The Chicago school of ethnography," in *Handbook of Ethnography*, ed. Paul Atkinson et al. (London: Sage, 2001): 11–25, quotes 20.

60. Blackman, "Ethnographic mosaic" (n. 56), 197–200.

61. Platt, "'Participant observation' method" (n. 2), quote 383.

62. Palmer, *Field Studies in Sociology* (n. 54), 48–55; and Bulmer, *Chicago School* (n. 4), 79–80. Also Ian Shaw, "Sociology and social work: An unresolved legacy of the Chicago School," in Hart, *Legacy of the Chicago School* (n. 54), 44–64; Shaw (44) notes that the divorce was "welcomed to some extent on both sides."

63. Palmer, *Field Studies in Sociology* (n. 54), quote 50.

64. Palmer, *Field Studies in Sociology* (n. 54), 168–79 (method of interview). Also Stuart A. Rice, "Hypotheses and verification in Clifford R. Shaw's studies of

juvenile delinquency," in *Methods in Social Science*, ed. Rice (Chicago: University of Chicago Press, 1931), 549–65; and Gieryn, "City as truth-spot" (n. 47), 23–25.

65. Ernest W. Burgess, "What social case records should contain to be useful for sociological interpretations," *Social Forces* 6 (1928): 524–32, at 524–25, quote 527.

66. Bulmer, *Chicago School* (n. 4), 102–4.

67. Clifford R. Shaw, *Jack-Roller* (n. 55), 22–23. Jack-rollers were specialists who robbed drunks.

68. On case study: Platt, "Case study" (n. 58), 17–23, 37; Palmer, *Field Studies in Sociology* (n. 54), 19–22; and Wood, "Place of the community" (n. 17), 20–23. Also Charles H. Cooley, "Case study of small institutions as a method of research," in Cooley, *Sociological Theory and Social Research* (New York: Henry Holt, 1930), 313–22.

69. Frederic M. Thrasher, "The study of the total situation," *Journal of Educational Sociology* 1 (1928): 477–91. Also Clifford R. Shaw, "Case study method," *Publication of the American Sociological Society* 21 (1926): 149–57. Shaw's study describes the case of a delinquent boy who was always in trouble yet showed no signs of personal pathology. Study of the boy's total situation revealed that he was unable to reconcile the conflicting mores and expectations of his authoritarian, Greek-born parents and his free-wheeling Irish-American and German-American peers. The cause of the boy's unruly behavior was not anything in him, but the conflicting social and moral situation that he was unable to resolve.

70. Cooley, "Case study" (n. 68), 316.

71. Burgess, "What social case records" (n. 65), quote 527.

72. Quoted by Platt, "'Participant observation' method" (n. 2), 392n69.

73. Palmer, *Field Studies in Sociology* (n. 54), 36, 161–67.

74. Palmer, *Field Studies in Sociology* (n. 54), 36.

75. Palmer, *Field Studies in Sociology* (n. 54), 8–9.

76. Palmer, *Field Studies in Sociology* (n. 54), 104–5, 107, quote 104.

77. Frederic M. Thrasher, "How to study the boys' gang in the open," *Journal of Educational Sociology* 1 (1928): 244–54; and Thrasher, *The Gang: A Study of 1,313 Gangs in Chicago* (Chicago: University of Chicago Press, 1927). Also Blackman, "Ethnographic mosaic" (n. 56), 208–11. Blackman calls Thrasher's work "ethnographic."

78. Platt, "'Participant observation' method" (n. 2), 383; and Bulmer, *Chicago School* (n. 4), 104.

79. Paul G. Cressey, "A comparison of the roles of the 'sociological stranger' and the 'anonymous stranger' in field research," *Urban Life* 12 (1983): 102–20. Cressey wrote this prescient essay around 1927, two years into his project. Martin Bulmer, "The methodology of the *Taxi-Dance Hall*: An early account of Chicago ethnography from the 1920s," *Urban Life* 12 (1983): 95–101; and Platt, "'Participant observation' method" (n. 2), 382–83.

80. Fred B. Lindstrom and Ronald A. Hardert, eds., "Kimball Young on founders of the Chicago school," *Sociological Perspectives* 31 (1988): 269–97, at 280–81.

81. Lindstrom and Hardert, "Kimball Young on founders" (n. 80), 281.

82. Pauline V. Young, *Scientific Surveys and Research* (New York: Prentice-Hall, 1939), quote 119–26 (Park's advice); and Young, *The Pilgrims of Russian Town* (Chicago: University of Chicago Press, 1932).

83. Ernest W. Burgess, "Preface," in Albert B. Blumenthal, *Small-Town Stuff* (Chicago: University of Chicago Press, 1932), ix–xvi, quotes xi and x.

84. Burgess, "Preface" (n. 83), quotes x–xi. Also Harvey, *Myths of the Chicago School* (n. 48), 64–66, 65 (aerial mapping).

85. Blumenthal to Ernest Burgess, 20 April 1929, box 3, file 1, Burgess Papers, Special Collections, Regenstein Library, University of Chicago, quoted in Harvey, *Myths of the Chicago School* (n. 48), 65, 66. Blumenthal did not publish the full life histories, lest intimate revelations of identifiable people bring small-town wrath upon the sociologist and his family. For the same reason he did not publish a methodological chapter on his practice of intimate conversation. Blumenthal, *Small-Town Stuff* (n. 83), 147. Including these would have made the book more obviously ethnographic.

86. Blumenthal, *Small-Town Stuff* (n. 83), chaps. 6–10.

87. Faris, *Chicago Sociology* (n. 4), 66.

88. Nels Anderson, "A stranger at the gate: Reflections on the Chicago school of sociology," *Urban Life* 11 (1983): 396–406, at 403–4.

89. Nels Anderson, *The Hobo: The Sociology of the Homeless Man* (Chicago: University of Chicago Press, 1923; Phoenix edition, 1961), 3–15, quote 14–15. All subsequent references to *Hobo* are to the 1961 edition.

90. Anderson, *Hobo* (n. 89), 27–39 (institutions), 105–6 (migrant laborers), 109–21 (labor exchanges).

91. Anderson, *Hobo* (n. 89), 87–95, quote 87 from a book by Dr. Ben Reitman; 95 (class mobility), 96–106 (occupations of "home guard"), 105–6 (five main types), 40–57 ("getting by"). On the larger demographic and economic context, in which hobos performed indispensable pioneer labor: Frank Tobias Higbie, *Indispensable Outcasts: Hobo Workers and Community in the American Midwest, 1880–1930* (Urbana: University of Illinois Press, 2003).

92. Nels Anderson, "The private and last frontier," *Sewanee Review* 77 (1969): 25–90, quote 69.

93. Nels Anderson, *American Hobo: An Autobiography* (Leiden, Netherlands: Brill, 1975), 2–28 (early life), 29–36 (Traverse Bay); and Anderson, *Hobo* (n. 89), v–xxi, vii (family moves).

94. Anderson, *American Hobo* (n. 93), 8–9.

95. Anderson, *American Hobo* (n. 93), 23–28.

96. Anderson, *American Hobo* (n. 93), 36–41, quote 142.

97. Anderson, *American Hobo* (n. 93), 86; and Anderson, *Hobo* (n. 89), v–xxi, vii (siblings leave home).

98. Anderson, *American Hobo* (n. 93), 40–51, 60, quote 49.

99. The following is drawn from Anderson, *American Hobo* (n. 93).

100. Anderson, "Stranger at the gate" (n. 88), 396–98.

101. Anderson, *Hobo* (n. 89), xi.

102. Anderson, *American Hobo* (n. 93), 157–59, quote 157.

103. Interview of G. B. Johnson by James Carey, 27 Mar. 1972, in University of Chicago Regenstein Library, Special Collections, quoted in Harvey, *Myths of the Chicago School* (n. 48), 58. Johnson spent occasional weekends with Anderson in Hobohemia, where they rented a hotel room and walked around talking to people.

104. Anderson, *American Hobo* (n. 93), 160–61, 164; and Anderson, "Stranger at the gate" (n. 88), 397, 401.

105. Anderson, "Stranger at the gate" (n. 88), 399; and Anderson, *American Hobo* (n. 93), 161.

106. Anderson, *American Hobo* (n. 93), 163–66 (work and study interfere); and Anderson, "Stranger at the gate" (n. 88), 400 (Mugsy). Also Ernest W. Burgess, "Committee's preface," in Anderson, *Hobo* (n. 89), xxvii–xxix, at xxvii; and Platt, " 'Participant observation' method" (n. 2), 382 (survey of 400 hobos).

107. Anderson, *American Hobo* (n. 93), 162–65; and Anderson, "Stranger at the gate" (n. 88), 400–401.

108. Anderson, *American Hobo* (n. 93), 162, 165 (quote); Anderson, *Hobo* (n. 89), xi; and Anderson, "Stranger at the gate" (n. 88), 400.

109. Anderson, *American Hobo* (n. 93), 166.

110. Anderson, *Hobo* (n. 89), xi–xii, quotes xii.

111. Anderson, *Hobo* (n. 89), 85–86; and Burgess, "Committee's preface" (n. 106), quote xxviii.

112. Anderson, *American Hobo* (n. 93), 164.

113. Anderson, *Hobo* (n. 89), chaps. 1–4. This part included an account of the edge-of-town "jungles" that served hobos as way stations between road and main stem. In these out-of-the-way places hobos could wash, cook or find a meal, and teach and reinforce the moral economy of hobo social life.

114. Anderson, *Hobo* (n. 89), chaps. 9–17 and appendix B.

115. Anderson, *Hobo* (n. 89), 30–33.

116. Anderson, *Hobo* (n. 89), 61–95, quote 61.

117. Anderson, *American Hobo* (n. 93), 163–66.

118. Anderson, *American Hobo* (n. 93), 19–20, quote 20.

119. Burgess, "Committee's preface" (n. 106), xxvii–xviii.

120. Anderson, *American Hobo* (n. 93), 165–68, quote 165–66.

121. Anderson, *American Hobo* (n. 93), 166.

122. These generalizations are based on material and ideas in Anderson, *Hobo* (n. 89), chap. 5, 61–86.

123. Anderson, *Hobo* (n. 89), 109.

124. Anderson, *Hobo* (n. 89), quote 92.

125. Anderson, *American Hobo* (n. 93), quote 52.

126. Anderson, *American Hobo* (n. 93), 23–24, quote 23.

127. Anderson, *Hobo* (n. 89), quote viii–ix.

128. Anderson, *Hobo* (n. 89), quote ix.

129. Anderson, *American Hobo* (n. 93), 66.

130. Anderson, *American Hobo* (n. 93), 131 (Brigham Young University); and Anderson, "Stranger at the gate" (n. 88), 401–3 (University of Chicago), quote 402.

131. Anderson, "Stranger at the gate" (n. 88), quote 397.

132. Anderson, *American Hobo* (n. 93), quote 157; and Anderson, "Stranger at the gate" (n. 88), 396 (courses). Anderson had taken courses in sociology thinking they would be an easy route to a college degree and a possible background for a career in law.

133. Young, *Scientific Surveys* (n. 82), 122–23, quote 123.

134. Anderson, *Hobo* (n. 89), xiii.

135. Platt, "'Participant observation' method" (n. 2), 382.

136. Anderson, *Hobo* (n. 89), quote xiii.

137. Anderson, "Stranger at the gate" (n. 88), 404; and Anderson, *Hobo* (n. 89), 169.

138. Anderson, *Hobo* (n. 89), xii.

139. Nels Anderson, "The Social Antecedents of the Slum: A Developmental Study of the East Harlem Area of Manhattan Island, New York City," PhD dissertation, New York University, 1930.

140. Anderson, *American Hobo* (n. 93), 171–83.

CHAPTER FOUR

1. William F. Whyte, *Street Corner Society: The Social Structure of an Italian Slum* (Chicago: University of Chicago Press, 1943; reissued in 1955 with appendix, and in 1981 with expanded appendix). All citations are to the 4th edition, 1993. Also Whyte, *Participant Observer: An Autobiography* (Ithaca, NY: IRL Press Cornell University, 1994).

2. Jennifer Platt, "The development of the 'participant observation' method in sociology: Origin myth and history," *Journal of the History of the Behavioral Sciences* 19 (1983): 379–93, at 385–87, quote 385. Also Platt, *A History of Sociological Research Methods in America, 1920–1960* (Cambridge: Cambridge University Press, 1996), 251.

3. Clifford Wilcox, *Robert Redfield and the Development of American Anthropology* (Lanham, MD: Lexington Books, 2004), 47.

4. Luther L. Bernard, "Development of methods in sociology," *Monist* 38 (1928): 292–320, at 308–10, 318; and Jesse Bernard, "The history and prospects of sociology in the United States," in *Trends in American Sociology*, ed. George A. Lundberg, Read Bain, and Nels Anderson (New York: Harper, 1929), 1–71, esp. 46–47, 67–68.

5. On border zones: Robert E. Kohler, *Landscapes and Labscapes: Exploring the Lab-Field Border in Biology* (Chicago: University of Chicago Press, 2002), chap. 1.

6. Ralph L. Beals, "Acculturation," in *Anthropology Today: An Encyclopedic Inventory*, ed. Alfred L. Kroeber (Chicago: University of Chicago Press, 1953), 621–41; and Melville J. Herskovits, *Acculturation: The Study of Culture* (New York: J. J. Augustin, 1938).

7. Bronislaw Malinowski, "Practical anthropology," *Africa* 2 (1929): 22–38, at 36–37, quote 36; George W. Stocking, Jr., *After Tylor: British Social Anthropology, 1888–1951* (Madison: University of Wisconsin Press, 1995), chap. 8; and Henrika Kuklick, *The Savage Within: The Social History of British Anthropology, 1885–1945* (New York: Cambridge University Press, 1991), chaps. 5–6.

8. Robert Redfield, Ralph Linton, and Melville J. Herskovits, "Memorandum for the study of acculturation," *American Anthropologist* 38 (1936): 149–52; and Beals, "Acculturation" (n. 6), 622–23.

9. Herskovits, *Acculturation* (n. 6).

10. James West (pseud. for Carl Withers), *Plainville, U.S.A.* (New York: Columbia University Press, 1945); and Carl Withers, "Foreword," in Art Gallaher, *Plainville Fifteen Years Later* (New York: Columbia University Press, 1961), vii–ix.

11. Jesse Bernard, "History and prospects" (n. 4), quote 68; Clark Wissler, *Man and Culture* (New York: Thomas Crowell, 1923); and Alfred L. Kroeber, *Anthropology* (New York: Harcourt Brace, 1923). On the influence of anthropologists' concept of culture on social science: John S. Gilkeson, *Anthropologists and the Rediscovery of America, 1886–1965* (New York: Cambridge University Press, 2010). Gilkeson deals narrowly with students of Franz Boas and concepts of democracy and class. I am grateful to Lynn Nyhart for calling my attention to this work.

12. Robert E. Park, "The city: Suggestions for the investigation of human behavior in the urban environment," in *The City*, ed. Ernest W. Burgess and Roderick D. McKenzie (Chicago: University of Chicago Press, 1925; reprinted 1967), 1–46, at 3.

13. Arthur E. Wood, "The place of the community in sociological studies," *Publication of the American Sociological Society* 22 (1928): 14–25, at 20–23.

14. On community study: Jesse F. Steiner, "The sources and methods of community study," in *The Fields and Methods of Sociology*, ed. Luther L. Bernard (New York: Ray Long and Richard Smith, 1934), 303–12, esp. 307–9. Also Arthur J. Vidich, Joseph Bensman, and Maurice R. Stein, eds., *Reflections on Community Studies* (New York: Wiley, 1964). Bibliographical sources include W. Lloyd Warner, "Social anthropology and the modern community," *American Journal of Sociology* 46 (1941): 785–96; and Julian Steward, "Area research, theory and practice," *Social Science Research Council Bulletin* 63 (1950): 1–164.

15. Everett C. Hughes, *French Canada in Transition* (Chicago: University of Chicago Press, 1943); John Dollard, *Caste and Class in a Southern Town* (New Haven, CT: Yale University Press, 1937); and August Hollingshead, *Elmtown's Youth: The Impact of Social Classes on Adolescents* (New York: Wiley, 1949). Hughes and Dollard were Chicago trained (PhDs 1928 and 1931); Hollingshead's study was a postdoctoral project of University of Chicago's Committee on Human Development.

16. Everett C. Hughes, "The place of fieldwork in the social sciences," in *Field-*

work: An Introduction to the Social Sciences, ed. Buford H. Junker (Chicago: University of Chicago Press, 1962), v–xv, quote x.

17. Martin Bulmer, Kevin Bales, and Kathryn Kish Sklar, eds., *The Social Survey in Historical Perspective* (Cambridge: Cambridge University Press, 1991); Jean M. Converse, *Survey Research in the United States: Roots and Emergence, 1890–1960* (Berkeley: University of California Press, 1987); Sarah E. Igo, *The Averaged American: Surveys, Citizens, and the Making of a Mass Public* (Cambridge, MA: Harvard University Press, 2007); and Platt, *History of Sociological Research Methods* (n. 2).

18. Robert S. Lynd and Helen Merrell Lynd, *Middletown: A Study in Contemporary American Culture* (New York: Harcourt Brace, 1929), 3–4.

19. Richard W. Fox, "Epitaph for Middletown: Robert S. Lynd and the analysis of consumer culture," in *The Culture of Consumption: Critical Essays in American History, 1880–1980*, ed. Fox and T. Jackson Lears (New York: Pantheon, 1983), 101–41. Also Gilkeson, *Anthropologists and the Rediscovery* (n. 11), 69–90.

20. Lynd and Lynd, *Middletown* (n. 18), 505–10, quotes 506 and 507.

21. Platt, "'Participant observation' method" (n. 2), 384, quote 391n30.

22. Although *Middletown*'s immense popularity earned Robert Lynd a PhD and a professorship in sociology at Columbia, his academic situation proved uncongenial, and he gradually reverted to his original avocation of social critic and activist. He left two ambitious studies of urban middle-class life unfinished, and a follow-up study of Muncie proved an anticlimax. Fox, "Epitaph" (n. 19), 132–33, 140–41.

23. George W. Stocking, Jr., *Anthropology at Chicago: Tradition, Discipline, Department* (Chicago: Joseph Regenstein Library, 1979), 11–17, quote 21.

24. Wilcox, *Robert Redfield* (n. 3), 15, 27–29; Stocking, *Anthropology at Chicago* (n. 23), 23–25; and Gilkeson, *Anthropologists and the Rediscovery* (n. 11), 218–37, esp. 219–20.

25. Wilcox, *Robert Redfield* (n. 3), 29–32; Gilkeson, *Anthropologists and the Rediscovery* (n. 11), 220–22 (not winning confidence); and Robert Redfield, *Tepoztlán, a Mexican Village: A Study of Folk Life* (Chicago: University of Chicago Press, 1930).

26. Wilcox, *Redfield* (n. 3), 41–48.

27. The Yucatán survey was an add-on to an archaeological project whose leader, Alfred Kidder, expected Redfield to catalogue Indian and Spanish elements in visual art for use in historical archaeology. Redfield, however, saw an opportunity to further his interest in acculturation, and with Park's backing he brought Kidder around to his plan. Sociologists generally liked the results; anthropologists markedly less so. It was their project, after all, that had been hijacked. Wilcox, *Redfield* (n. 3), 49–60; Stocking, *Anthropology at Chicago* (n. 23), 23; and Robert Redfield, *The Folk Culture of Yucatan* (Chicago: University of Chicago Press, 1941).

28. Horace M. Miner, *St. Denis: A French-Canadian Parish* (Chicago: University of Chicago Press, 1939).

29. Hughes, *French Canada* (n. 15); Everett C. Hughes, "French Canada: The natural history of a research project," in Vidich et al., *Reflections on Community*

Studies (n. 14), 71–83; and Howard S. Becker, *What about Mozart? What about Murder? Reasoning from Cases* (Chicago: University of Chicago Press, 2014), 31 (on Hughes).

30. Hughes, "Place of fieldwork" (n. 16), quote vi (ideas and lives); and David Riesman and Howard S. Becker, "Introduction to the Transaction edition," in Everett C. Hughes, *The Sociological Eye: Selected Papers* (New Brunswick, NJ: Transaction, 1993), v–xiv, ix (proud of forebears).

31. Hughes, "Place of fieldwork" (n. 16), xi–xii, quote xii.

32. Gary A. Fine, ed., *A Second Chicago School? The Development of a Postwar American Sociology* (Chicago: University of Chicago Press, 1995); Jennifer Platt, "Research methods and the second Chicago school," in Fine, *Second Chicago School*, 82–107; David Riesman, "The legacy of Everett Hughes," *Contemporary Sociology* 12 (1983): 477–81; and Lewis A. Coser, "Introduction," in *Everett C. Hughes: On Work, Race, and the Sociological Imagination*, ed. Coser (Chicago: University of Chicago Press, 1994), 1–17.

33. On Becker: Adam Gopnik, "The outsider game: How the sociologist Howard Becker studies the conventions of the unconventional," *New Yorker*, 12 Jan. 2015, 26–31, quote 27; "Dialogue with Howard S. Becker (1970): An interview conducted by Julius Debro," in Howard S. Becker, *Doing Things Together: Selected Papers* (Evanston, IL: Northwestern University Press, 1986), 25–46, at 25–26; "Howard Becker and Alain Mueller discuss *Telling about Society*," howardsbecker.com/articles/Suisse%20English.html, 10; and Ken Plummer, "Continuity and change in Howard S. Becker's work: An interview with Howard S. Becker," *Sociological Perspectives* 46 (2003): 21–39.

34. Herbert J. Gans, "The participant-observer as a human being: Observations on the personal aspects of fieldwork," in *Institutions and the Person*, ed. Howard W. Becker, Blanche Geer, David Riesman, and Robert S. Weiss (Chicago: Aldine, 1968), 300–17, at 300–301.

35. Lambros Comitas, "Conrad Maynadier Arensberg (1910–1997)," *American Anthropologist* 101 (1999): 810–13, quote 811; Conrad M. Arensberg and Solon T. Kimball, *Family and Community in Ireland* (Cambridge, MA: Harvard University Press, 1940); Arensberg, *The Irish Countryman: An Anthropological Study* (New York: Macmillan, 1937).

36. Richard Gillespie, *Manufacturing Knowledge: A History of the Hawthorne Experiments* (New York: Cambridge University Press, 1991), esp. chaps. 4–6; Fritz J. Roethlisberger, *The Elusive Phenomena: An Autobiographical Account of My Work in the Field of Organizational Behavior at the Harvard Business School* (Cambridge, MA: Harvard University Press, 1977), esp. 54–57; and George Caspar Homans, *Coming to My Senses* (New Brunswick, NJ: Transaction, 1984), 134–66 (Mayo at Harvard).

37. Gillespie, *Manufacturing Knowledge* (n. 36), quote 259, 155. Mayo assigned Malinowski and Radcliffe-Brown to his seminars but was interested in anthropology

only as it illuminated contemporary society. Homans, *Coming to My Senses* (n. 36), 153–55.

38. Gillespie, *Manufacturing Knowledge* (n. 36), 152–57, quote 156 (Mayo).

39. Gillespie, *Manufacturing Knowledge* (n. 36), 152–57; Warner, "Social anthropology" (n. 14), 785–96; and Gilkeson, *Anthropologists and the Rediscovery* (n. 11), 90–117. Gilkeson writes (94) that Newburyport was in fact not socially homogeneous. Nearly half its population were ethnic minorities; it was only the dominant social elites that were homogeneously old Yankee.

40. Lloyd Warner, "Preface," in Arensberg and Kimball, *Family and Community* (n. 35), xi–xiv. Also Allison Davis, Burleigh B. Gardner, and Mary R. Gardner, *Deep South: A Social Anthropology of Caste and Class* (Chicago: University of Chicago Press, 1941).

41. Whyte, *Participant Observer* (n. 1), 45–52.

42. Whyte, *Participant Observer* (n. 1), 25–35.

43. Whyte, *Participant Observer* (n. 1), quote 51.

44. Caroline F. Ware, *Greenwich Village: A Comment on American Civilization in the Post-war Years* (Boston: Houghton Mifflin, 1935), esp. "Note on method and sources," 427–35 (appendix A). Ware began her study as a settlement-house social survey and resorted to unstructured interview and observing only when standard questionnaires failed to work with local residents—a familiar pattern. On Ware's understudied life, Wikipedia is a useful starting point.

45. Whyte, *Participant Observer* (n. 1), 62–65; and Whyte, *Street Corner Society* (n. 1), 361, 284–88. Whyte's recollections in these two books differ in significant details. Whyte seemed uninterested in Chicago sociologists' neighborhood studies; why is not clear.

46. Whyte, *Street Corner Society* (n. 1), xi, 286–88.

47. Whyte, *Participant Observer* (n. 1), 62–66.

48. William F. Whyte, "The social role of the settlement house," *Applied Anthropology* 1 (1941): 14–19, esp. 17.

49. Whyte, *Participant Observer* (n. 1), 65–66; and Whyte, *Street Corner Society* (n. 1), 288–90.

50. Whyte, *Participant Observer* (n. 1), 67–69; and Whyte, *Street Corner Society* (n. 1), 291–93, 298–300, quote 299.

51. Whyte, *Street Corner Society* (n. 1), 8–14, 255–58.

52. Whyte, *Street Corner Society* (n. 1), 298–301, quote 307; and Whyte, *Participant Observer* (n. 1), 72–73.

53. Whyte, *Street Corner Society* (n. 1), 293–98, 302–7; and Whyte, *Participant Observer* (n. 1), 69–72.

54. Whyte, *Street Corner Society* (n. 1), quote 279.

55. Whyte, *Street Corner Society* (n. 1), 291–93, 298–300, quote 299.

56. Whyte, *Street Corner Society* (n. 1), 302–4, quote 303.

57. Whyte, *Street Society* (n. 1), xv–xvi; and Whyte, "Social role of the settlement house" (n. 48).

58. Whyte, *Street Corner Society* (n. 1), 301; and Whyte, *Participant Observer* (n. 1), 73–74.

59. Whyte, *Street Corner Society* (n. 1), 303.

60. Whyte, *Street Corner Society* (n. 1), 286–88; and Whyte, *Participant Observer* (n. 1), 53–57, 263–64. Also Eliot D. Chapple with Conrad M. Arensberg, *Measuring Human Relations: An Introduction to the Study of the Interaction of Individuals*, Genetic Psychology Monograph 22 (Provincetown, MA: Journal Press, 1940). Chapple was driven to invent his sociometric machine by his failure to derive an objective picture of Newburyport's class structure from the particularistic data of individual interviews. Gilkeson, *Anthropologists and the Rediscovery* (n. 11), 97. Chapple and Arensberg were also influenced by the positivistic "operationalist" philosophy of the Harvard physicist Percy Bridgman, which was much discussed in Henderson's circle at the time. Homans, *Coming to My Senses* (n. 36), 162–63; and William F. Whyte, *Learning from the Field: A Guide from Experience* (Newbury Park, CA: Sage, 1984), 63–64. Also Joel Isaac, *Working Knowledge: Making the Human Sciences from Parsons to Kuhn* (Chicago: University of Chicago Press, 2012), chap. 2.

61. William F. Whyte, "On *Street Corner Society*," in *Contributions to Urban Sociology*, ed. Ernest W. Burgess and Donald J. Bogue (Chicago: University of Chicago Press, 1964), 256–68, quote 262.

62. Hughes, "Place of fieldwork" (n. 16), quotes xiv and xv.

63. Whyte, *Participant Observer* (n. 1), quote 96.

64. Whyte, *Street Corner Society* (n. 1), 52–78, 94–98, 104–8, 320–25; and Whyte, *Participant Observer* (n. 1), 97–98.

65. Whyte, *Learning from the Field* (n. 60), quote 23.

66. Whyte, *Street Corner Society* (n. 1), 255–62.

67. Whyte, *Street Corner Society* (n. 1), 318.

68. Whyte, *Street Corner Society* (n. 1), 317–19; and Whyte, *Participant Observer* (n. 1), 83–84.

69. Whyte, *Street Corner Society* (n. 1), 14–25, 2; and Whyte, *Participant Observer* (n. 1), 83–84.

70. Whyte, *Street Corner Society* (n. 1), 319–20, quote 320.

71. Whyte, *Street Corner Society* (n. 1), quote 318–19.

72. Whyte, *Participant Observer* (n. 1), 83–84.

73. Whyte, *Street Corner Society* (n. 1), 45–49, 265–66, 327–28, quotes 328. Also Whyte, *Participant Observer* (n. 1), 99–100.

74. Whyte, *Street Corner Society* (n. 1), 35–44, 265–68; and Whyte, *Participant Observer* (n. 1), 98–100.

75. Allan R. Holmberg, "Experimental intervention in the field," in *Peasants, Power, and Applied Social Change: Vicos as a Model*, ed. Henry F. Dobyns, Paul L. Doughty, and Harold Lasswell (London: Sage, 1971), 21–32, at 28–30, quotes 28.

76. Whyte, *Street Corner Society* (n. 1), 309–17, also chap. 6.

77. Whyte, *Street Corner Society* (n. 1), 328–36, also chap. 4.

78. Whyte, *Street Corner Society* (n. 1), 262–63, 328–36, 367–68.

79. On case study: John Forrester, "If *p*, then what? Thinking in cases," *History of the Human Sciences* 9 (1996): 1–25; Charles C. Ragin and Howard S. Becker, eds., *What Is a Case? Exploring the Foundations of Social Inquiry* (New York: Cambridge University Press, 1992); Mary S. Morgan, "Case studies: One observation or many? Justification or discovery?," *Philosophy of Science* 79 (2012): 667–77, at 672–73; and Morgan, "Peeling open the case study," unpublished paper, Princeton University Davis Center, 2011, 13–26.

80. Whyte, *Street Corner Society* (n. 1), 279–80, quote 280.

81. Whyte, *Street Corner Society* (n. 1), 322–23, quote 323.

82. Gans, "Participant-observer," (n. 34), quote 309.

83. Whyte, *Street Corner Society* (n. 1), 354–57; and Whyte, *Participant Observer* (n. 1), 138.

84. On the postwar hegemony of scientistic polling survey: George Steinmetz, "American sociology before and after World War II: The (temporary) settling of a disciplinary field," in *Sociology in America: A History*, ed. Craig Calhoun (Chicago: University of Chicago Press, 2007), 314–66. Ethnographic community studies include Herbert J. Gans, *The Urban Villagers: Group and Class in the Life of Italian-Americans* (New York: Free Press, 1962); Gans, *The Levittowners: Ways of Life and Politics in a New Suburban Community* (New York: Pantheon, 1967); Hollingshead, *Elmtown's Youth* (n. 15); John R. Seeley, R. Alexander Sim, and Elizabeth W. Loosley, *Crestwood Heights: A Study in the Culture of Suburban Life* (New York: Basic Books, 1956); Arthur J. Vidich and Joseph Bensman, *Small Town in Mass Society: Class, Power, and Religion in a Rural Community* (Garden City, NY: Doubleday, 1958); and Elliot Liebow, *Talley's Corner: A Study of Negro Streetcorner Men* (Boston: Little Brown, 1967).

85. Samuel A. Stouffer, "Some observations on study design," *American Journal of Sociology* 55 (1950): 355–61, quote 359. In his dissertation Stouffer claimed to have proven by rigorous comparative analysis that random sampling and questionnaire were better in every respect than observing in situ. In fact, he demonstrated only that polling survey was better for the one purpose of measuring opinions on single issues without regard to social situations—the sole purpose for which polling survey had been designed. Stouffer, "An Experimental Comparison of Statistical and Case History Methods of Attitude Research," PhD dissertation, University of Chicago, 1930.

86. Stouffer, "Some observations on study design" (n. 85), 359, 360.

87. Whyte, *Street Corner Society* (n. 1), xv–xvii, quote xvi, 272–76.

88. Whyte, *Participant Observer* (n. 1), 108–15. Whyte was allowed to satisfy the requirement of a literature review by publishing one as a separate article. The University of Chicago Press accepted *Street Corner Society* only with a substantial author's subsidy.

89. Gans, "Participant-observer" (n. 34), 316–17. "Alienation" was the overused trope of Cold War culture and social science.

90. Hortense Powdermaker, *Stranger and Friend: The Way of an Anthropologist* (London: Secker and Warburg, 1967); and Herbert J. Gans, "Working in six research areas: A multi-field sociological career," *Annual Review of Sociology* 35 (2009): 1–19.

CHAPTER FIVE

1. Dale Peterson, *Jane Goodall: The Woman Who Redefined Man* (Boston: Houghton Mifflin, 2006).

2. Peterson, *Jane Goodall* (n. 1), quote 227. On the history of field primatology: Pamela J. Asquith, "The intellectual history of field studies in primatology," in *Strength in Diversity: A Reader in Physical Anthropology*, ed. Ann Herring and Leslie Chan (Toronto: Canadian Scholar's Press, 1994), 49–75; Amanda Rees, "A place that answers questions: Primatological field sites and the making of authentic observations," *Studies in History and Philosophy of Biology and Biomedical Sciences* 37 (2006): 311–33; Rees, *The Infanticide Controversy: Primatology and the Art of Field Science* (Chicago: University of Chicago Press, 2009), chaps. 1–2; and Georgina M. Montgomery, *Primates in the Real World: Escaping Primate Folklore and Creating Primate Science* (Charlottesville: University of Virginia Press, 2015). Montgomery associates Goodall with the "primate folklore" of "pop primatology." Also Donna J. Haraway, *Primate Visions: Gender, Race, and Nature in the World of Modern Science* (New York: Routledge, 1989), together with Haraway's second thoughts about the thesis of her celebrated book, in Haraway, *When Species Meet (Posthumanities)* (Minneapolis: University of Minnesota Press, 2007), 312–13n29.

3. Stuart A. Altman, "Preface," in *Social Communication among Primates*, ed. Altman (Chicago: University of Chicago Press, 1967), ix–xii, at xi (figure 1); Irven DeVore and Richard Lee, "Recent and current field studies of primates," *Folia Primatologica* 1 (1963): 66–72; and Jane Goodall, "Cultural elements in a chimpanzee community," in *Precultural Primate Behavior*, vol. 1, ed. Emil W. Menzel (Basel, Switzerland: Karger, 1973), 144–84, at 162, 153.

4. Altman, "Preface" (n. 3), xi–xii; Asquith, "Intellectual history" (n. 2), quote 65 (psychologists); William Mason, "Naturalistic and experimental investigations of the social behavior of monkeys and apes," in *Primates: Studies in Adaptation and Variability*, ed. Phyllis C. Jay (New York: Holt, Rinehart, and Winston, 1968), 398–419, at 401–3; and Thelma E. Rowell, "The myth of peculiar primates," in *Mammalian Social Learning: Comparative and Ecological Perspectives*, ed. Hilary O. Box and Kathleen R. Gibson (Cambridge: Cambridge University Press, 1999), 6–16.

5. George B. Schaller, *The Mountain Gorilla: Ecology and Behavior* (Chicago: University of Chicago Press, 1963); and Schaller, *The Year of the Gorilla* (Chicago: University of Chicago Press, 1964).

6. Pamela J. Asquith, "Primate research groups in Japan: Orientations and east-west differences," in *The Monkeys of Arashiyama: Thirty-Five Years of Research in*

Japan and the West, ed. Linda M. Fedigan and Asquith (Albany, NY: SUNY Press, 1991), 81–98; Asquith and Arne Kaland, eds., *Japanese Images of Nature: Cultural Perspectives* (Richmond, UK: Curzon Press, 1997); and Kinji Imanishi, "Social organization of subhuman primates in their natural habitat," *Current Anthropology* 1 (1960): 393–407.

7. Vernon Reynolds and Frances Reynolds, "Chimpanzees in the Budongo Forest," in *Primate Behavior: Field Studies of Monkeys and Apes*, ed. Irven DeVore (New York: Holt, Rinehart, and Winston, 1965), 368–424; and Vernon Reynolds, *Budongo: An African Forest and Its Chimpanzees* (Garden City, NY: Natural History Press, 1965).

8. Shirley C. Strum and Linda M. Fedigan, *Primate Encounters: Models of Science, Gender, and Society* (Chicago: University of Chicago Press, 2000).

9. Jane Goodall, *In the Shadow of Man* (Boston: Houghton Mifflin, 1971), 13–16, quote 14; and Goodall, *Reason for Hope: A Spiritual Journey* (New York: Warner, 1999), 61.

10. On the Gombe reserve: Jane Goodall, *The Chimpanzees of Gombe: Patterns of Behavior* (Cambridge, MA: Belknap Press of Harvard University Press, 1986), 43–51, map 46; Peterson, *Jane Goodall* (n. 1), 117–21, 179–84; and Goodall, *Africa in My Blood: An Autobiography in Letters*, ed. Dale Peterson (Boston: Houghton Mifflin, 2000), esp. Peterson's introduction, 152–54, and Goodall to family, 16 July 1960, 157–58.

11. Jane Goodall, "The behavior of free-living chimpanzees in the Gombe Stream area," *Animal Behavior Monographs* 1, no. 3 (1968): 161–311, at 165.

12. Peterson, *Jane Goodall* (n. 1), 38–60 (childhood), 79–83, 87–98 (Clo and Africa).

13. Peterson, *Jane Goodall* (n. 1), 53–60 (schooling), 58 (career adviser), 69 (Vanne's advice), 69–71 (secretarial school). Jane entered secretarial school in 1953 and got her certificate in 1954. Her examiners disapproved of her "childish fantasy" of becoming a journalist but thought she would eventually settle down in a proper secretarial job (71).

14. Sources on Leakey include Sonia M. Cole, *Leakey's Luck: The Life of Louis Seymour Bazett Leakey, 1903–1972* (New York: Harcourt Brace Jovanovich, 1975), esp. chap. 12; Peterson, *Jane Goodall* (n. 1), 98–100 and passim; Louis S. B. Leakey, *White African: An Early Autobiography* (London: Hodder, 1937); Louis S. B. Leakey, *By the Evidence: Memoirs, 1932–1951* (New York: Harcourt Brace Jovanovich, 1974); and Mary D. Leakey, *Disclosing the Past* (New York: Doubleday, 1984).

15. On Tigoni: Cole, *Leakey's Luck* (n. 14), 325–33. The primatologist Irven DeVore observed, perhaps unfairly, that what Leakey knew about animal behavior was "the stereotype stuff you get from white hunters at a bar in Nairobi." Peterson, *Jane Goodall* (n. 1), 220.

16. Peterson, *Jane Goodall* (n. 1), 116–20, 103–4 (Rosalie Osborne); and Sy Montgomery, *Walking with the Great Apes: Jane Goodall, Dian Fossey, Biruté Galdikas* (Boston: Houghton Mifflin, 1991).

17. Peterson, *Jane Goodall* (n. 1), quote 146.

18. The following is drawn from Peterson, *Jane Goodall* (n. 1), 100–121, 146–55 (funding); and Goodall, *Africa* (n. 10), 93–103, 114.

19. Peterson, *Jane Goodall* (n. 1), 98–121; Goodall, *Africa* (n. 10), 93–103, esp. Goodall to family, 30 May and 3 or 4 July 1957; and Goodall, *Reason for Hope* (n. 9), 44–46, 52–58.

20. Peterson, *Jane Goodall* (n. 1), 119–21, 145–46 (Vanne), 152–55; Goodall to family, early Sept. 1957, in Goodall, *Africa* (n. 10), 114.

21. Goodall to family, probably 19 Sept. 1960, in Goodall, *Africa* (n. 10), 159–62, quote 162.

22. Peterson, *Jane Goodall* (n. 1), 58 (pet photographer), 97–98, 107 (sportswoman), 151–52 (zoo), 115–18, 122–25 (museum); and Goodall, *Reason for Hope* (n. 9), 53–58 (Coryndon).

23. Useful sources on field methods in primatology include George B. Schaller, "Field procedures," in DeVore, *Primate Behavior* (n. 7), 623–29; and Rees, *Infanticide Controversy* (n. 2).

24. Peterson, *Jane Goodall* (n. 1), 167–72. Lolui Island was one of the Leakeys' favorite research sites.

25. Peterson, *Jane Goodall* (n. 1), 183–88; and Goodall, *Shadow of Man* (n. 9), 19–21.

26. Peterson, *Jane Goodall* (n. 1), 187–88, 238–39, 194 (decision to concentrate). Comments on the haphazardness of forest work are found throughout Goodall's letters; for example, Goodall to family 16 July, 30 Aug., 19 Sept. 1960, in Goodall, *Africa* (n. 10), 157–61.

27. Goodall, *Shadow of Man* (n. 9), 24–32, is the most vivid account of her finding the Peak. Also Peterson, *Jane Goodall* (n. 1), 194–96; Goodall to family, 30 Aug. 1960, in Goodall, *Africa* (n. 10), 158–59; and Goodall, *Reason for Hope* (n. 9), 63–65. In September Goodall replaced her minders with expert trackers on loan from a friendly Kenyan hunter-planter. Peterson, *Jane Goodall* (n. 1), 197.

28. Goodall to family, June 1963, Goodall, *Africa* (n. 10), 251–57.

29. Goodall, *Shadow of Man* (n. 9), 30.

30. Peterson, *Jane Goodall* (n. 1), 197.

31. Goodall to family, 31 Oct. / 1 Nov. 1960, in Goodall, *Africa* (n. 10), 163–66, quote 164; Goodall to family, May 1966, 362–63 (killing and eating); and Peterson, *Jane Goodall* (n. 1), 205–7. Also Goodall, *Chimpanzees of Gombe* (n. 10), 296–312, 372–74.

32. Goodall, *Shadow of Man* (n. 9), 35–37; Goodall, *Reason for Hope* (n. 9), 66–67; Peterson, *Jane Goodall* (n. 1), 207–11; and Goodall, *Chimpanzees of Gombe* (n. 10), 248–51, 258–61, 536–39.

33. Peterson, *Jane Goodall* (n. 1), 202–5, quote 205. The Schallers were impressed by Goodall's determination to succeed and ability to climb hills fast; also by the short time she had been given to make good, and her poor equipment (second-rate binoculars, bad camera, no telescope).

34. Peterson, *Jane Goodall* (n. 1), 212–14, quote 214 (important to science).

35. Goodall, *Shadow of Man* (n. 9), 57–61, 203–4, quote 61.

36. Goodall to Uncle Eric, 13 July 1961, in Goodall, *Africa* (n. 10), 184–85, quote 184.

37. Goodall to family, 14 July 1961, in Goodall, *Africa* (n. 10), 185–89, quote 186.

38. Goodall to family, 6 Feb. 1961, in Goodall, *Africa* (n. 10), 167–75, quote 172; Goodall, *Shadow of Man* (n. 9), 52–54; and Peterson, *Jane Goodall* (n. 1), 234–37, quoting Goodall's detailed field note of the rain-dance episode, 235.

39. Edgar Rice Burroughs, *Tarzan of the Apes* (Oxford: Oxford University Press, 2010), 53–54. My thanks to Etienne Benson for directing me to this text.

40. Peterson says as much when he writes that "Gombe's magical world" was opened not by magic but by the "mundane keys" of hard, empirical fieldwork. Peterson, *Jane Goodall* (n. 1), 236.

41. Accounts of the development of provisioning include Peterson, *Jane Goodall* (n. 1), 238–44; Goodall, *Shadow of Man* (n. 9), 65–67; and Goodall to family, 16 May 1961, in Goodall, *Africa* (n. 10), 180–84. The fullest technical account is Richard W. Wrangham, "Artificial feeding of chimpanzees and baboons in the natural habitat," *Animal Behavior* 22 (1974): 83–93. On provisioning generally, see Pamela J. Asquith, "Provisioning and the study of free-ranging primates: History, effects, and prospects," *Yearbook of Physical Anthropology* 32 (1989): 129–58.

42. Goodall to Melvin Payne, 17 June 1963, in Goodall, *Africa* (n. 10), 250–52, quote 250.

43. Peterson, *Jane Goodall* (n. 1), 304–6; Goodall to family, late May / early June 1963, in Goodall, *Africa* (n. 10), 248–50; and Goodall, *Shadow of Man* (n. 9), 89–90, and chaps. 7 and 9 (Flo and family).

44. Peterson, *Jane Goodall* (n. 1), quote 323 (banana tree); and Goodall, *Shadow of Man* (n. 9), quote 94 (new camp).

45. Goodall to Bernard Verdcourt, 5 July 1963, in Goodall, *Africa* (n. 10), 258–61; and Peterson, *Jane Goodall* (n. 1), 245–61, 295–312.

46. Goodall, *Africa* (n. 10), 258–61, quote 260; and Peterson, *Jane Goodall* (n. 1), 337–39.

47. Goodall to Payne, 9 July 1964, in Goodall, *Africa* (n. 10), 298–304, quote 303.

48. Goodall to family, 21 Aug. 1964, in Goodall, *Africa* (n. 10), 310–15, quote 314.

49. Robert E. Kohler, *Lords of the Fly:* Drosophila *Genetics and the Experimental Life* (Chicago: University of Chicago Press, 1994), 46–49, 125–67.

50. Jane Goodall, "A preliminary report on expressive movements and communication in free-ranging chimpanzees," in *Primates: Studies in Adaptation and Variability*, ed. Phyllis Jay (New York: Holt, Rinehart, and Winston, 1968), 313–74, at 337–39; and Goodall, *Shadow of Man* (n. 9), 89–90.

51. Goodall to "FFF" [Louis Leakey], 30 Aug. 1963, in Goodall, *Africa* (n. 10), 263–66, quote 264; and Peterson, *Jane Goodall* (n. 1), 304–6, 345–47. "FFF" is short for Foster Fairy Father.

52. On the move from Lake to Ridge Camp: Peterson, *Jane Goodall* (n. 1), 348–53; Goodall, *Africa* (n. 10), 284–98; and Goodall, *Shadow of Man* (n. 9), chap. 8, esp. 94–95.

53. The following account is drawn from Wrangham, "Artificial feeding" (n. 41), esp. 84–85; Goodall, *Shadow of Man* (n. 9), 140–45; and Peterson, *Jane Goodall* (n. 1), 348–53. Wrangham distinguishes five distinct regimens.

54. Goodall to family, May 1966, in Goodall, *Africa* (n. 10), 358–62, quote 359 (David Greybeard).

55. Goodall, *Shadow of Man* (n. 9), quote 140.

56. Wrangham, "Artificial feeding" (n. 41), 85–87, quote 87.

57. Peterson, *Jane Goodall* (n. 1), 322–323, quote 323.

58. Goodall, "Cultural elements" (n. 3), 162.

59. Goodall to family, 17 Aug. 1962, in Goodall, *Africa* (n. 10), 217–22, esp. 220; Goodall to Leonard Carmichael, 21 Oct. 1962, 226–28, esp. 227; and Goodall to Melvin Payne, 9 July 1964, 298–304, esp. 302. Also Strum and Fedigan, *Primate Encounters* (n. 8), 22–24.

60. Goodall "Expressive movements" (n. 50); Goodall, *Chimpanzees of Gombe* (n. 10), 114–45 and passim; and Jane Goodall, *My Friends the Wild Chimpanzees* (Washington, DC: National Geographic Society, 1967), 138–40.

61. Goodall, *Chimpanzees of Gombe* (n. 10), 207–30 (wandering and territory), 488–534 (warfare), 283–84 (cannibalism); and Jane Goodall, "Infant killing and cannibalism in free-living chimpanzees," *Folia Primatologica* 28 (1977): 259–82.

62. Goodall, *Shadow of Man* (n. 9), 120–21, quote 120.

63. Goodall, *Shadow of Man* (n. 9), 67–69; Goodall, "Expressive movements" (n. 50), 313–15; Goodall, "Cultural elements" (n. 3), 146–47; and Goodall, *Chimpanzees of Gombe* (n. 10), 146–206, 185–93 (chimps' social networks). Also Jane Goodall, "Chimpanzees of the Gombe Stream Reserve," in DeVore, *Primate Behavior* (n. 7), 451–57.

64. The following is drawn from Goodall, *Chimpanzees of Gombe* (n. 10), 418–24, 427–29, 431–35, and passim; and Goodall, *Shadow of Man* (n. 9), 121–26, 262–65, and passim.

65. Peterson, *Jane Goodall* (n. 1), 337.

66. Goodall, *Shadow of Man* (n. 9), 99.

67. Goodall to family, 17 Aug. 1962, in Goodall, *Africa* (n. 10), 217–22, esp. 220–21.

68. Goodall to Melvin Payne, 12 Oct. 1963, in Goodall, *Africa* (n. 10), 266–67.

69. Peterson, *Jane Goodall* (n. 1), 336.

70. Goodall to family 25 Apr. 1963, in Goodall, *Africa* (n. 10), 245–47, quote 247.

71. Peterson, *Jane Goodall* (n. 1), 226–28, quote 227. Kortlandt practiced a kind of outdoor experimental psychology, placing foreign objects (stuffed or tethered live animals, alien foods, man-made objects) where chimps would encounter them, and recording their reactions. Adriaan Kortlandt, "Experimentation with chimpanzees in

the wild," in *Neue Ergebnisse der Primatologie / Progress in Primatology* (Stuttgart, Germany: Gustav Fischer, 1967), 208–24.

72. Jane Goodall, *Through a Window: My Thirty Years with the Chimpanzees of Gombe* (Boston: Houghton Mifflin, 1990),17–18, 23.

73. F. Fraser Darling, *A Herd of Red Deer: A Study in Animal Behaviour* (Oxford: Oxford University Press, 1937; reprint, London: Geoffrey Cumberlege, 1941), quote 202.

74. Sy Montgomery, *Walking with the Great Apes* (n. 16), 269–73; Dian Fossey, *Gorillas in the Mist* (Boston: Houghton Mifflin, 1983); and Georgina M. Montgomery, *Primates in the Real World* (n. 2), chap. 5. On anthropomorphism: Pamela J. Asquith, "Anthropomorphism and the Japanese and Western traditions in primatology East and West," in *Primate Ontogeny, Cognition, and Behavior: Developments in Field and Laboratory Research*, ed. J. Else and P. Lee (New York: Academic, 1986), 61–71; Robert W. Mitchell, Nicholas S. Thompson, and H. Lyn Miles, eds., *Anthropomorphism, Anecdotes, and Animals* (Albany, NY: SUNY Press, 1997); Lorraine Daston and Gregg Mitman, eds., *Thinking with Animals: New Perspectives on Anthropomorphism* (New York: Columbia University Press, 2005); and Etienne Benson, "Naming the ethological subject," *Science in Context* 29 (2016): 107–28. In the 1980s anthropomorphism became a more acceptable way of understanding animals: William A. Mason, "Can primate political traits be identified?" in Else and P. Lee, *Primate Ontogeny*, 3–11, esp. 3–4.

75. Goodall, *My Friends* (n. 60), 191–92.

76. Sy Montgomery, *Walking with the Great Apes* (n. 16), 268–69, quote 269.

77. Peterson, *Jane Goodall* (n. 1), 322–24, quote 323.

78. Peterson, *Jane Goodall* (n. 1), 170–72, quote 170–71.

79. Peterson, *Jane Goodall* (n. 1), 38–42.

80. Peterson, *Jane Goodall* (n. 1), 40–42, 105; Goodall, *Africa* (n. 10), 95–103; and Goodall, *Reason for Hope* (n. 9), 42–43.

81. Dale Peterson, section introduction in Goodall, *Africa* (n. 10), 231–235, at 231–32.

82. Peterson, *Jane Goodall* (n. 1), quote 83 (attuned), 29–52, esp. 32–33 (family circle).

83. Peterson, *Jane Goodall* (n. 1), 94–95, quote 94.

84. Peterson, *Jane Goodall* (n. 1), 122–23, 142, 145–46, quotes 121 (Jane) and 145 (Vanne), 123–25, 130 (Leakey).

85. Goodall to family, late Jan. 1965, in Goodall, *Africa* (n. 10), 339–45, quote 341; Peterson, *Jane Goodall* (n. 1), 365–68. For a photograph of the three alpha males with Goodall: Cole, *Leakey's Luck* (n. 14), 336.

86. Thelma E. Rowell, "A few peculiar primates," in Strum and Fedigan, *Primate Encounters* (n. 8), 57–70, at 65–68.

87. Peterson, *Jane Goodall* (n. 1), quote 598; Goodall to family, probably 19 Sept. 1960, in Goodall, *Africa* (n. 10), 159–61, quote 160; Goodall, *Shadow of Man* (n. 9), quote 131 (Hugo).

88. Charles Elton, *Animal Ecology* (London: Sidgwick and Jackson, 1927), quote 64.

89. Goodall to family, 17 Aug. 1962, in Goodall, *Africa* (n. 10), 217–22, quote 221.

90. Goodall, *My Friends* (n. 60); Peterson, *Jane Goodall* (n. 1), 393–94, 480–81. Goodall was bemused by her storybook alter ego but accepted it as the price of the National Geographic Society's support—a useful, if unsought, celebrity. Goodall to Melvin Payne, 9 July 1964, in Goodall, *Africa* (n. 10), 298–304.

91. Goodall, *Chimpanzees of Gombe* (n. 10); and Peterson, *Jane Goodall* (n. 1), 337, 437, 441–42. The two types of data were kept as separate series in their own record books. Peterson, *Jane Goodall* (n. 1), 437.

92. Peterson, *Jane Goodall* (n. 1), quote 481; and Goodall, *Shadow of Man* (n. 9).

93. Goodall to Huxley, 16 Sept. 1971, quoted in Greg Mitman, "Life in the field: The sensuous body as popular naturalist's guide," in Strum and Fedigan, *Primate Encounters* (n. 8), 421–35, at 433. The letter is in box 44, folder 1, Julian Sorell Huxley Papers, Woodson Research Center, Fondrieu Library, Rice University, Houston, TX.

94. Peterson, *Jane Goodall* (n. 1), in a letter home 3 Aug. 1962, quote 281.

95. Peterson, *Jane Goodall* (n. 1), 281–82, 284.

96. Peterson, *Jane Goodall* (n. 1), 284–94; and Rowell, "Few peculiar primates" (n. 86), 58–59. Also Solly Zuckerman, "Concluding remarks," *Symposium of the Zoological Society of London* 10 (1963): 119–23.

97. Peterson, *Jane Goodall* (n. 1), 285–86, quotes 286 (Napier) and 290 (Morris). John Napier had tutored Goodall informally when she was in London waiting to go to Gombe.

98. Goodall, *Through a Window* (n. 72), 14–15.

99. Peterson, *Jane Goodall* (n. 1), 261–63, 271–77, 284–85; and Robert A. Hinde, "Some reflections on primatology at Cambridge and the science studies debate," in Strum and Fedigan, *Primate Encounters* (n. 8), 104–15.

100. Mason, "Naturalistic and experimental" (n. 4), 398.

101. Sy Montgomery, *Walking with the Great Apes* (n. 16), 105–6, quote 106.

102. Peterson, *Jane Goodall* (n. 1), 273–77; Sy Montgomery, *Walking with the Great Apes* (n. 16), 105–6; and Goodall, *Through a Window* (n. 72), 14–16.

103. Sy Montgomery, *Walking with the Great Apes* (n. 16), quote 106.

104. Peterson, *Jane Goodall* (n. 1), 275–77; Goodall, *Through a Window* (n. 72), quotes 15 (truculent) and 16 (trappings).

105. Hinde, "Some reflections on primatology" (n. 99), 113 (learning from Goodall); and Goodall, *Through a Window* (n. 72), 15 (editing). On ethologists' eventual accommodation to naming and narrating: Benson, "Naming the ethological subject" (n. 74).

CHAPTER SIX

1. This chapter is a reworking of Robert E. Kohler, "Paul Errington, Aldo Leopold, and wildlife ecology: Residential science," *Historical Studies in the Natural Sciences* 41 (2011): 216–54.

2. Thomas R. Dunlap, *Saving America's Wildlife* (Princeton, NJ: Princeton University Press, 1988), chaps. 3–6; and Christian C. Young, *In the Absence of Predators: Conservation and Controversy on the Kaibab Plateau* (Lincoln: University of Nebraska Press, 2002). The Bureau of Biological Survey later became the US Fish and Wildlife Service.

3. Aldo Leopold, *Game Management* (New York: Charles Scribner's Sons, 1933); Leopold, *The River of the Mother of God and Other Essays*, ed. Susan L. Flader and J. Baird Callicott (Madison: University of Wisconsin Press, 1991); and Leopold, *For the Health of the Land: Previously Unpublished Essays and Other Writings*, ed. Callicott and Eric T. Freyfogle (Washington, DC: Island Press, 1999).

4. Sources include Dunlap, *Saving America's Wildlife* (n. 2); John F. Reiger, *American Sportsmen and the Origins of Conservation* (New York: Winchester Press, 1975); and Bonnie Christensen, "From divine nature to umbrella species: The development of wildlife science in the United States," in *Forest and Wildlife Science in America: A History*, ed. Harold K. Steen (Ashville, NC: Forest History Society, 1999), 209–29.

5. Lynn K. Nyhart, "*Wissenschaft* and *Kunde*: The general and the special in modern science," *Osiris* 27 (2012): 250–75.

6. The following paragraphs are drawn from Herbert L. Stoddard, *The Bobwhite Quail: Its Habits, Preservation, and Increase* (New York: Scribner's, 1931), esp. chaps. 2–3; Albert Way, *Conserving Southern Longleaf: Herbert Stoddard and the Rise of Ecological Land Management* (Athens: University of Georgia Press, 2011), 39–55; and Paul Errington and Frances N. Hamerstrom, Jr., "The northern bobwhite's winter territory," *Iowa Agricultural Experiment Station Research Bulletin* 201 (1936): 301–443, at 333–34.

7. Stoddard, *Bobwhite Quail* (n. 6), 349–58; and Way, *Conserving Southern Longleaf* (n. 6), 39.

8. Stoddard, *Bobwhite Quail* (n. 6), quote xxi.

9. Way, *Conserving Southern Longleaf* (n. 6), 7–14; and Lawrence S. Early, *Looking for Longleaf: The Fall and Rise of an American Forest* (Chapel Hill: University of North Carolina Press, 2004).

10. Way, *Conserving Southern Longleaf* (n. 6), 19–55, esp. 19–38. On "inner frontiers": Robert E. Kohler, *All Creatures: Naturalists, Collectors, and Biodiversity, 1850–1950* (Princeton, NJ: Princeton University Press, 2006), chap. 1.

11. On Baldwin: Way, *Conserving Southern Longleaf* (n. 6), 76–78; Herbert L. Stoddard, *Memoirs of a Naturalist* (Norman: University of Oklahoma Press, 1969), 165–66; S. Charles Kendeigh, "In memoriam: Samuel Prentiss Baldwin," *Auk* 57 (1940): 1–12; and S. Prentiss Baldwin, "Bird banding by means of systematic trapping," *Abstract of the Proceedings of the Linnaean Society of New York*, no. 31 (Dec. 1919): 23–56. Also Etienne Benson, "A centrifuge of calculation: Managing data and enthusiasm in early twentieth-century bird banding," *Osiris* 32 (2017): 286–306.

12. Way, *Conserving Southern Longleaf* (n. 6), 85–86; and Stoddard, *Memoirs* (n. 11), 165–67, 174–77, 184 (banding at Inwood).

13. Stoddard, *Memoirs* (n. 11), chaps. 1–3 (Florida), chaps. 4–5 (Wisconsin), chaps. 6–7 (museums).

14. Stoddard, *Memoirs* (n. 11), 176–77.

15. Kohler, *All Creatures* (n. 10), chaps. 2, 5; and Susan Leigh Star, "Craft vs. commodity, mess vs. transcendence: How the right tool became the wrong one in the case of taxidermy and natural history," in *The Right Tools for the Job: At Work in Twentieth-Century Life Sciences*, ed. Adele E. Clarke and Joan H. Fujimura (Princeton, NJ: Princeton University Press, 1992), 257–86.

16. Stoddard, *Memoirs* (n. 11), 223–24 (embarrassment); and Way, *Conserving Southern Longleaf* (n. 6), 62–72, quote 57.

17. For details of the project's organization: Stoddard, *Bobwhite Quail* (n. 6), xxi–xxix; and Way, *Conserving Southern Longleaf* (n. 6), 76–80. Stoddard reported to Waldo McAtee, the head of the Biological Survey's Food Habits Research division.

18. Stoddard, *Bobwhite Quail* (n. 6), xxiii.

19. Stoddard, *Bobwhite Quail* (n. 6).

20. Stoddard, *Bobwhite Quail* (n. 6), xxi; and Way, *Conserving Southern Longleaf* (n. 6), 78, 80.

21. Way, *Conserving Southern Longleaf* (n. 6), 80.

22. Stoddard, *Bobwhite Quail* (n. 6), quote 49–50.

23. Way, *Conserving Southern Longleaf* (n. 6), 90–95.

24. Way, *Conserving Southern Longleaf* (n. 6), 162–69 (patch farming), 119–21 (hunting customs), 92–115 (fire).

25. Way, *Conserving Southern Longleaf* (n. 6), quote 83.

26. Way, *Conserving Southern Longleaf* (n. 6); Leon Neel with Paul S. Sutter and Albert Way, *The Art of Managing Longleaf: A Personal History of the Stoddard-Neel Approach* (Athens: University of Georgia Press, 2010); and Stoddard, *Memoirs* (n. 11), chaps. 16–17.

27. Stoddard, *Bobwhite Quail* (n. 6), 15–19, 115, quote 19.

28. Stoddard, *Bobwhite Quail* (n. 6), 26–29 (egg laying), 29–36 (incubating), 36–40 (hatching), 40–47 (rearing, brooding, feeding), quote 31.

29. Stoddard, *Bobwhite Quail* (n. 6), 183–201. Most breeding pairs did eventually raise young, by double or triple nesting.

30. Stoddard, *Bobwhite Quail* (n. 6), 183–221.

31. Stoddard, *Bobwhite Quail* (n. 6), 113–57. The food-preference work was mainly done by an employee of the Biological Survey, Charles Handley, according to standard Survey protocols. Stoddard did not have a high opinion of the work.

32. Stoddard, *Memoirs* (n. 11), quote 177.

33. Way, *Conserving Southern Longleaf* (n. 6), quote 56–57.

34. Stoddard, *Memoirs* (n. 11), chaps. 1–3.

35. Stoddard, *Memoirs* (n. 11), 11–12 (fire), 29–58 (Florida years), quote 34.

36. Stoddard, *Memoirs* (n. 11), quote 58.

37. Way, *Conserving Southern Longleaf* (n. 6), quote 61.

38. Stoddard, *Memoirs* (n. 11), chaps. 4–5.

39. Stoddard, *Memoirs* (n. 11), 69–86, 75 (Ochsner's life), quote 77.

40. Stoddard, *Memoirs* (n. 11), 108–13, 145–53.

41. Stoddard, *Memoirs* (n. 11), 71–72 (the Wagners), 99–103, esp. 102–103 (the Laws), quote 102.

42. Stoddard, *Memoirs* (n. 11), quote 221.

43. Stoddard, *Memoirs* (n. 11), 154–63, 177; and Way, *Conserving Southern Longleaf* (n. 6), 73–75.

44. Stoddard, *Memoirs* (n. 11), quote 177.

45. Stoddard, *Memoirs* (n. 11), 230–37, quote 234, 237–51 (professional work at Sherwood); and Way, *Conserving Southern Longleaf* (n. 6), 143–53. Stoddard never did write another full life history, only a short guide to wild turkey habitat. Herbert L. Stoddard, *Maintenance and Increase of the Eastern Wild Turkey on Private Lands of the Coastal Plain of the Deep Southeast* (Tallahassee, FL: Tall Timbers Research Station, 1963).

46. Curt Meine, *Aldo Leopold: His Life and Work* (Madison: University of Wisconsin Press, 1988), 256–60, 264–69; and Aldo Leopold, *Report on a Game Survey of the North Central States* (Madison, WI: Democrat Press, 1931). The Depression ended SAAMI support for game surveys and research in 1931, but Leopold persuaded state governments to fund the fellowship program.

47. Stoddard, *Memoirs* (n. 11), 217–26; Way, *Conserving Southern Longleaf* (n. 6), 121–23; and Meine, *Aldo Leopold* (n. 46), 264–69.

48. Stoddard, *Memoirs* (n. 11), 219–22; and Paul Errington, *Predation and Life* (Ames: Iowa State University Press, 1967), 65.

49. Errington, *Predation and Life* (n. 48), quote 68.

50. Leopold to E. B. Fred, 8 Sept. 1938; and Aldo Leopold, "Why and how research," manuscript, Mar. 1948; both in box 3, folder 5, Paul L. Errington Papers (hereafter PLE), Iowa State University Archives, Ames.

51. The best overall sources on Errington's project are Paul Errington, "Some contributions of a fifteen-year local study of the northern bobwhite to a knowledge of population phenomena," *Ecological Monographs* 15 (1945): 1–34; and Errington, "Predation and vertebrate populations," *Quarterly Review of Biology* 21 (1946): 144–77, 221–45, at 161–63.

52. Quote in Errington to C. Boone, 2 July 1936, box 7, folder 3, PLE. Errington was referring to his quail study site near Ames, Iowa.

53. Errington to Gardiner Bump, 17 Sept. 1937; and Errington to Leopold, 18 Mar. 1937 (Gastrow); both in box 7, folder 5, PLE. Also Errington, *Predation and Life* (n. 48), 67–68.

54. Quote in Errington to Frederick Baumgartner, 3 Feb. 1939, box 8, folder 2, PLE.

55. Errington and Hamerstrom, "Northern bob-white's winter" (n. 6), fig. 23, at 394–95.

56. Sources on field techniques include Errington and Hamerstrom, "Northern bob-white's winter" (n. 6), 310–28; Paul Errington, "The wintering of the Wisconsin bobwhite," *Transactions of the Wisconsin Academy of Science, Arts, and Letters* 28 (1933): 1–35; and Errington, *Predation and Life* (n. 48), chaps. 5–6.

57. Errington and Hamerstrom, "Northern bob-white's winter" (n. 6), 328–31; Paul Errington, "The pellet analysis method of raptor food habits study," *Condor* 32 (1930): 292–96; Errington to Werner O. Nagel, 18 Jan. 1939, box 8, folder 1, PLE; Errington, *Predation and Life* (n. 48), 3–5; and Errington, "Wintering " (n. 56), 9–10 (poachers).

58. Errington to Waldo L. McAtee, 28 Nov. 1934, box 7, folder 2, PLE; and Stoddard to Errington, 30 Nov., 16 Dec. 1929, 27 Mar. 1930, box 5, folder 5, PLE.

59. Errington to Leopold, 18 Mar. 1937, box 7, folder 5, PLE; 14 Sept. 1943, box 8, folder 6, PLE; and Leopold to Errington, 4 Sept. 1943, box 3, folder 5, PLE.

60. Errington and Hamerstrom, "Northern bob-white's winter" (n. 6), quotes 317 and 310.

61. Paul Errington, "Vulnerability of bob-white populations to predation," *Ecology* 15 (1934): 110–27, at 112, 122–24.

62. Range managers invented the term and concept of carrying capacity to justify regulation of overgrazing by cattle, and these were then extended to deer and other ungulates—hence their narrow emphasis on the single factor of food supply as limiting factor. The concept became more inclusive when it was adopted by ecologists, thanks in part to Errington's work on cover. R. Y. Edwards and C. David Fowle, "The concept of carrying capacity," *Transactions of the 20th Wildlife Management Conference* (Washington, DC: Wildlife Management Institute, 1955), 589–602; and Christian C. Young, "Defining the range: The development of carrying capacity in management practice," *Journal of the History of Biology* 31 (1998): 61–83.

63. Errington, "Predation and vertebrate populations" (n. 51), quote 226.

64. On extermination campaigns: Dunlap, *Saving America's Wildlife* (n. 2); and Young, *In the Absence of Predators* (n. 2).

65. Timothy Fridtjof Flannery, *The Future Eaters: An Ecological History of the Australasian Lands and People* (New York: Braziller, 1995), 93–94 (exterminator species).

66. Errington initially used "carrying capacity" and "thresholds of security" interchangeably, but as his views evolved he gave up the confusing loan word for his own neologism.

67. The detailed history of Errington's field results is compiled from Errington and Hamerstrom, "Northern bob-white's winter" (n. 6), 361–62; Errington, "Some contributions" (n. 51), 11–12; and Paul Errington, "The Northern Bobwhite: Environmental Factors Influencing Its Status," PhD dissertation, University of Wisconsin, 1932, chap. 10. Also Errington to Waldo L. McAtee, 29 July 1932, 18 May 1933; Errington to S. Charles Kendeigh, 27 Jan. 1933; and Errington to Herbert L. Stoddard, 10 May, 29 Sept. 1933; all in box 7, folder 1, PLE (pages 2 and 3 of the 18 May letter are misfiled in box 4, folder 1).

68. Quote from Errington to Stoddard, 29 Sept. 1933; and Errington to William T. Hornaday, 9 May 1933; both in box 7, folder 1, PLE.

69. Errington, *Predation and Life* (n. 48), quote 87.

70. Summaries of field data: Errington, *Predation and Life* (n. 48), 92–99; and Errington, "Some contributions " (n. 51), 8–11.

71. Errington to Stoddard, 17 Jan. 1936, box 7, folder 4, PLE; Errington to Leopold, 29 Dec. 1942, box 8 folder 5, PLE; 30 Mar., 26 July 1943, box 8, folder 6, PLE; Errington and Hamerstrom, "Northern bob-white's winter" (n. 6), 394–98, 402; Paul Errington, "Eight-winter study of central Iowa bob-whites," *Wilson Bulletin* 53 (1941): 85–102, at 89–92, 98–100; Errington, "Predation and vertebrate populations" (n. 51), 161–63; Errington to Margaret Nice, 10 Jan., 7 May 1940, box 8, folder 3, PLE.

72. Errington to J. A. Ferguson, 7 Jan. 1936, box 7, folder 4, PLE.

73. Stoddard to Errington, 31 Dec. 1937, box 5, folder 5, PLE.

74. Carolyn Errington to the author, 8 June 2000.

75. Paul Errington, *The Red Gods Call* (Ames: Iowa State University Press, 1973), 3–7.

76. Kohler, *All Creatures* (n. 10), chaps. 1–2; Paul Errington, *Of Men and Marshes* (Ames: Iowa State University Press, 1957), 116–17, 125–29, and passim. Quote in Errington to Otto Koehler, 9 Jan. 1948, box 9, folder 1, PLE.

77. Errington, *Red Gods Call* (n. 75), chaps. 2–4 ("kid-hunter"); Errington, "Come-back from polio," undated typescript, box 3, folder 5, PLE; quote in J. N. Darling to Errington, 4 Jan. 1958, box 2, folder 1, PLE.

78. Errington, *Men and Marshes* (n. 76), quote 130.

79. Errington to Wallace B. Grange, 4 Jan. 1958, box 2, folder 1, PLE.

80. Errington, *Red Gods Call* (n. 75), 15, 102; and Errington, *Men and Marshes* (n. 76), quote 136.

81. Errington, *Red Gods Call* (n. 75), 37–47.

82. Errington, *Red Gods Call* (n. 75), chap. 10.

83. Errington, *Red Gods Call* (n. 75), chap. 11.

84. Errington to Wilfred Osgood, 15 Apr. 1923, box 3, folder E-20, Osgood Papers, Field Museum, Chicago. Errington was writing about a possible museum job studying animal behavior in the field.

85. Carolyn Errington to the author, 8 June 2000 (n. 74). Also Errington to Marty C. Hamaker, 13 Dec. 1934, box 7, folder 2, PLE.

86. Errington, *Red Gods Call* (n. 75), 153 (backsliding), quote 137.

87. Errington, *Red Gods Call* (n. 75), 102–9, quotes 105 (benumbed) and 108 (metamorphosis).

88. Errington, *Red Gods Call* (n. 75), quote 108–9.

89. Leopold's evolving philosophy can be followed in these essays in Leopold, *River of the Mother of God* (n. 3): "A biotic view of land [1939]," 266–73; "The outlook for farm wildlife [1945]," 323–26, 326 (nice to have around); and "The

ecological conscience [1947]," 338–46. Also Aldo Leopold, "Planning for wildlife [1941]," in *Health of the Land* (n. 3), 193–98, 197 (land as factory).

90. On residents as conservationists: Leopold, "The farmer as a conservationist [1939]," and "Land use and democracy [1942]," both in Leopold, *River of the Mother of God* (n. 3): 255–65, 295–300; Leopold "Be your own emperor [c. 1938]," in *Health of the Land* (n. 3), 70–81, quote 78–79 (localized scientific reasoning); and Leopold to Seth Gordon, 30 Mar. 1931, quoted in Meine, *Aldo Leopold* (n. 46), 280.

91. Julianne Lutz Newton, *Aldo Leopold's Odyssey* (Washington, DC: Island Press, 2006), 306–10; Meine, *Aldo Leopold* (n. 46), 340–42, 364–65, 374–76, 381–82, and passim. Also Aldo Leopold, *A Sand County Almanac and Sketches Here and There* (New York: Oxford University Press, 1949).

92. On Leopold's early career: Meine, *Aldo Leopold* (n. 46), parts 2–3; Susan L. Flader, *Thinking like a Mountain: Aldo Leopold and the Evolution of an Ecological Attitude toward Deer, Wolves, and Forests* (Madison: University of Wisconsin Press), chaps. 1–3; Newton, *Aldo Leopold's Odyssey* (n. 91); and J. Baird Callicott and Eric T. Freyfogle, "Introduction," in Leopold, *For the Health of the Land* (n. 3), 3–26.

93. Char Miller, *Gifford Pinchot and the Making of Modern Environmentalism* (Washington, DC: Island Press, 2001); and Brian Balogh, "Scientific forestry and the roots of the modern American state: Gifford Pinchot's path to progressive reform," *Environmental History* 7 (2002): 198–225.

94. Callicott and Freyfogle, "Introduction" (n. 92), 6–7, 16–18, quote 16–17. On second nature: William Cronon, *Nature's Metropolis: Chicago and the Great West* (New York: Norton, 1991), esp. xvii, 56–57, 61–63, 101.

95. Leopold, *Game Management* (n. 3), endnotes.

96. Quote from Leopold to Ovid Butler, 10 Oct. 1934, box 3, folder 5, PLE.

97. Leopold to Errington, 26 June [1933], box 3, folder 5, PLE. "[1937]" is penciled in, but internal evidence dates the letter unambiguously to 1933. Also Leopold to Errington, 20 Mar. 1933, box 3, folder 5, PLE.

98. Quote from Errington to Thomas G. Scott, 6 Feb. 1950, box 9, folder 3, PLE.

99. Quote from Errington to C. G. Snook, 3 Oct. 1932, box 7, folder 1, PLE.

100. Leopold, introduction to round-table discussion for the Ecological Society, 31 Dec. 1947, box 3, folder 5, PLE, quote 3.

101. Errington to Leopold, 31 Mar. 1937, box 7, folder 5, PLE; and Paul Errington, "Of population cycles and unknowns," *Cold Spring Harbor Symposia in Quantitative Biology* 22 (1957): 287–300, quote 287.

102. Leopold to Errington, 8 Jan. 1944, cited by Newton, *Aldo Leopold's Odyssey* (n. 91), 382; Leopold to Errington, 20 Feb. 1935, box 3, folder 5, PLE; and Errington to Leopold, 14 Oct. 1941, 14 Sept. 1943, box 8, folders 4 and 6, PLE.

103. Aldo Leopold, "The role of universities in game conservation," *DuPont Magazine* 25, no. 6 (June 1931): 8–9, 24, quoted in Way, *Conserving Southern Longleaf* (n. 6), 84.

104. Errington to Leopold, 26 July, 14 Sept., 15 Oct. 1943, box 8, folder 6, PLE; and Leopold to Errington, 12 July 1943 (withdraws), 4 Sept. 1943 (hypotheses), and 17 Sept. 1943; all in box 3, folder 5, PLE.

EPILOGUE

1. Bronislaw Malinowski, *Crime and Custom in Savage Society* (London: Kegan Paul, Trench, Trubner, 1926), 179–89.

2. Alfred W. Crosby, *Ecological Imperialism: The Biological Expansion of Europe, 900–1900* (New York: Cambridge University Press, 1986). On the concept of "second nature": William Cronon, *Nature's Metropolis: Chicago and the Great West* (New York: Norton, 1991), esp. xvii, 56–57, 101, and 267–69.

3. Richard W. Burckhardt, Jr., *Patterns of Behavior: Konrad Lorenz, Niko Tinbergen, and the Founding of Ethology* (Chicago: University of Chicago Press, 2005); and Marcia Myers Bonta, *Women in the Field: America's Pioneering Women Naturalists* (College Station: Texas A&M University Press, 1991), chaps. 19 (Sherman) and 21 (Nice).

4. Jonathan Weiner, *The Beak of the Finch: A Story of Evolution in Our Time* (New York: Knopf, 1994).

5. Charles Elton, *The Pattern of Animal Communities* (London: Methuen, 1966); and Peter Crowcroft, *Elton's Ecologists: A History of the Bureau of Animal Population* (Chicago: University of Chicago Press, 1991), 94–105.

6. Nick Jans, *The Grizzly Maze: Timothy Treadwell's Fatal Obsession with Alaskan Bears* (New York: Penguin, 2005).

7. Adolph Murie, *The Wolves of Mount McKinley*, Fauna Series, no. 5 (Washington, DC: National Park Service, 1944).

8. Etienne Benson, *Wired Wilderness: Technologies of Tracking and the Making of Modern Wildlife* (Baltimore, MD: Johns Hopkins University Press, 2010).

9. David Lowenthal, *The Past Is a Foreign Country* (Cambridge: Cambridge University Press, 1985; revised edition, 2015).

10. Martin J. S. Rudwick, *Scenes from Deep Time: Early Pictorial Representations of the Prehistoric World* (Chicago: University of Chicago Press, 1992); and Rudwick, "Encounters with Adam, or at least the hyenas: Nineteenth century representations of the deep past," in *History, Humanity and Evolution: Essays for John C. Greene*, ed. James R. Moore (Cambridge: Cambridge University Press, 1989), 231–52.

11. Sources for archaeology include Colin Renfrew, *Approaches to Social Archaeology* (Cambridge, MA: Harvard University Press, 1984); Gordon R. Willey and Jeremy A. Sabloff, *A History of American Archaeology* (third edition, New York: Freeman, 1993), chap. 6; David L. Clarke, ed., *Spatial Archaeology* (New York: Academic, 1977); and Renfrew, *Before Civilization: The Radiocarbon Revolution and Prehistoric Europe* (New York: Knopf, 1973). On biogeochemistry and deep time: Peter Ward and Joe Kirschvink, *A New History of Life: The Radical New Discoveries about the Origins and Evolution of Life on Earth* (New York: Bloomsbury, 2015).

12. Evelyn Fox Keller, *A Feeling for the Organism: The Life and Work of Barbara McClintock* (New York: W. H. Freeman, 1983), quote 207 (my emphasis). Also Nathaniel C. Comfort, *The Tangled Field: Barbara McClintock's Search for the Patterns of Genetic Control* (Cambridge, MA: Harvard University Press, 2001).

13. Alix Cooper, *Inventing the Indigenous: Local Knowledge and Natural History in Early Modern Europe* (New York: Cambridge University Press, 2007).

14. Jane Jacobs, *The Death and Life of Great American Cities* (New York: Random House, 1961). A good inside story of project science is Tracy Kidder, *The Soul of a New Machine* (Boston: Little Brown, 1991).

15. Philip J. Pauly, "Summer resort and scientific discipline," in *The American Development of Biology*, ed. Ronald Rainer, Keith R. Benson, and Jane Maienschein (Philadelphia: University of Pennsylvania Press, 1988), 121–50. Also Peder Anker, "Science as a vacation: A history of ecology in Norway," *History of Science* 45 (2007): 455–79.

16. James C. Scott, *Seeing like a State: How Certain Schemes to Improve the Human Condition Have Failed* (New Haven, CT: Yale University Press, 1998), chap. 8.

17. Susan D. Jones, workshop presentation, University of Pennsylvania, 2016; and Jones, email to author, 15 Sept. 2017. "Nomadic peoples," Jones writes, "seemed to see a large area as encompassing their 'homelands' but this included multiple locations (in the western way of thinking), a much more fluid and (dare I say) ecological concept of 'place.'"

INDEX